CLEP

College Level Examination Program

Mathematics Series

Copyright © 2016

All rights reserved. No part of the material protected by this copyright notice may be reproduced or utilized in any form or by any means, electronic or mechanical, including photocopying or recording or by any information storage and retrievable system, without written permission from the copyright holder.

To obtain permission(s) to use the material from this work for any purpose including workshops or seminars, please submit a written request to:

XAMonline, Inc.
21 Orient Avenue
Melrose, MA 02176
Toll Free: 1-800-301-4647
Email: info@xamonline.com
Web: www.xamonline.com
Fax: 1-617-583-5552

Library of Congress Cataloging-in-Publication Data
Wynne, Sharon

CLEP Mathematics Series/ Sharon Wynne
   ISBN: 978-1-60787-581-9

1. CLEP   2. Study Guides   3. Mathematics

Disclaimer:

The opinions expressed in this publication are the sole works of XAMonline and were created independently from the College Board, or other testing affiliates. Between the time of publication and printing, specific test standards as well as testing formats and website information may change that are not included in part or in whole within this product. XAMonline develops sample test questions, and they reflect similar content as on real tests; however, they are not former tests. XAMonline assembles content that aligns with test standards but makes no claims nor guarantees candidates a passing score.

Cover photos provided by © Can Stock Photo Inc./kmitu/0405102; © Can Stock Photo Inc./yurolaitsalbert/16472806; © Can Stock Photo Inc./vtorous/7098607; © Can Stock Photo Inc./airdone/26886042; © Can Stock Photo Inc./sdmix/5783896

Printed in the United States of America
CLEP Mathematics Series
ISBN: 978-1-60787-581-9

TABLE OF CONTENTS

College Math 1

    Sample Test 3

    Answer Key 17

    Rationales 18

College Algebra 50

    Sample Test 51

    Answer Key 63

    Rationales 64

PreCalculus 101

    Sample Test 103

    Answer Key 115

    Rationales 116

Calculus 149

    Sample Test 151

    Answer Key 161

    Rationales 163

COLLEGE MATH

Description of the Examination

The College Mathematics exam covers material generally taught in a college course for nonmathematics majors and majors in fields not requiring knowledge of advanced mathematics.

The examination contains approximately 60 questions to be answered in 90 minutes. Some of these are pretest questions that will not be scored. Any time test takers spend on tutorials and providing personal information is in addition to the actual testing time.

An online scientific (nongraphing) calculator will be available during the examination. Although a calculator is not necessary to answer most of the questions, there may be a few problems whose solutions are difficult to obtain without using a calculator. Since no calculator is allowed during the examination except for the online calculator provided, is it recommended that prior to the examination you become familiar with the use of the online calculator.

For more information about downloading the practice version of the scientific (nongraphing) calculator, please visit the College Mathematics description on the CLEP website, **clep.collegeboard.org** It is assumed that test takes are familiar with currently taught mathematics vocabulary, symbols, and notation.

Knowledge and Skills Required

Questions on the College Mathematics examination require test takers to demonstrate the following abilities in the approximate proportion indicated.

- Solving routine, straightforward problems (about 50% of the examination)

- Solving nonroutine problems requiring an understanding of concepts and the application of skills and concepts (about 50% of the examination)

The subject matter of the College Mathematics examination is drawn from the following topics. The percentages next to the main topics indicate the approximate percentage of exam questions on that topic.

20% **Algebra**
- Solving equations, linear inequalities, and systems of linear equations by analytical and graphical methods
- Interpretation, representation, and evaluation of functions: numerical, graphical, symbolic, and descriptive methods
- Graphs of functions: translations, horizontal and vertical reflections, and symmetry about the x-axis, the y-axis, and the origin
- Linear and exponential growth
- Applications

10% **Counting and Probability**
- Counting problems: the multiplication rule, combinations and permutations
- Probability: union, intersection, independent events, mutually exclusive events, complementary events, conditional probabilities, and expected value
- Applications

COLLEGE MATH

15% **Data Analysis and Statistics**
- Data interpretation and representation: tables, bar graphs, line graphs, circle graphs, pie charts, scatterplots, and histograms
- Numerical summaries of data: mean (average), median, mode, and range
- Standard deviation, normal distribution (conceptual questions only)
- Applications

20% **Financial Mathematics**
- Percents, percent change, markups, discounts, taxes, profit, and loss
- Interest: simple, compound, continuous interest, effective interest rate, effective annual yield or annual percentage rate (APR)
- Present value and future value
- Applications

10% **Geometry**
- Properties of triangles and quadrilaterals: perimeter, area, similarity, and the Pythagorean theorem
- Parallel and perpendicular lines
- Properties of circles: circumference, area, central angles, inscribed angles, and sectors
- Applications

15% **Logic and Sets**
- Logical operations and statements: conditional statements, conjunctions, disjunctions, negations, hypotheses, logical conclusions, converses, inverses, counterexamples, contrapositives, logical equivalence
- Set relationships, subsets, disjoint sets, equality of sets, and Venn diagrams
- Operations on sets: union, intersection, complement, and Cartesian product
- Applications

10% **Numbers**
- Properties of numbers and their operations: integers and rational, irrational, and real numbers (including recognizing rational and irrational numbers)
- Elementary number theory: factors and divisibility, primes and composites, odd and even integers, and the fundamental theorem of arithmetic
- Measurement: unit conversion, scientific notation, and numerical precision
- Absolute value
- Applications

COLLEGE MATH

SAMPLE TEST

DIRECTIONS: Read each item and select the best response.

1. Which of the following is closed under division?

    I. $\left\{\frac{1}{3}, 1, 3\right\}$

    II. $\{-1, 1\}$

    III. $\{-1, 0, 1\}$

    A. I only

    B. II only

    C. III only

    D. I and II

    E. II and III

2. Which of the following is always composite if x is an odd positive integer and y is an even positive integer greater than 1?

    A. $x + y$

    B. $|x + y|$

    C. $x + 2y$

    D. $3x + y$

    E. $3xy$

3. Find the LCM of 25, 18, and 24.

    A. 1200

    B. 1800

    C. 2400

    D. 3600

    E. 10,800

4. Solve for x: $|3x| + 6 = 21$

    A. $[9, -5]$

    B. $[-9, 5]$

    C. $[-5, 0, 5]$

    D. $[-5, 5]$

    E. $[-9, 9]$

5. Which graph represents the solution set for $x^2 - 5x > -6$?

    A. number line with open circles at -2 and 2

    B. number line with open circles at -3 and 3

    C. number line with open circles at -2 and 2

    D. number line with open circles at 2 and 3

    E. number line with open circles at 2 and 3

6. What is the equation of the graph shown below?

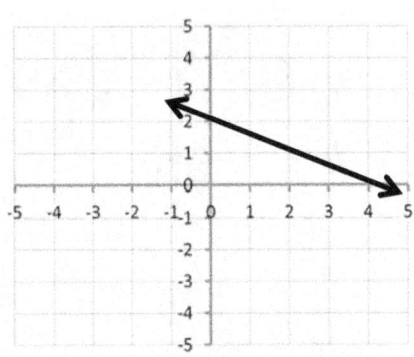

A. $x + 2y = 4$

B. $x - 2y = 4$

C. $2x + y = 4$

D. $x + 2y = -4$

E. $x - 2y = -4$

7. Solve the following inequality:
$-2x > 4$

A. $x > -2$

B. $x < -2$

C. $x > 2$

D. $x > -8$

E. $x < 2$

8. Which equation represents a circle centered on the origin with radius 3?

A. $x^2 + y^2 = 3$

B. $x^2 + y^2 = 6$

C. $x^2 + y^2 = 9$

D. $x^2 + y^2 = 36$

E. $x^2 - y^2 = 9$

9. Given that D is a distance, M is a mass, T is a time, and V is a velocity, which of the following units could be used to measure $\frac{MTV}{D}$?

A. feet

B. meters

C. grams

D. seconds

E. miles per hour

10. Cubic meters are used to measure which of the following?

A. Distance

B. Length

C. Area

D. Volume

E. Mass

11. What figure best describes a data set in which many items are clustered near the median value with a smaller number of values less than or greater than the median at greater distances on each side?

    A. A parabola

    B. A normal curve

    C. A line of best fit

    D. A Cartesian curve

    E. A Newtonian curve

12. If you prove a theorem by showing that an attempt to prove the opposite of the theorem leads to a contradiction, you are using the logical strategy called:

    A. Inductive reasoning

    B. Exhaustive proof

    C. Proof by attraction

    D. Direct proof

    E. Indirect proof

13. Compute the area of the shaded region, given a radius of 7 meters. Point O is the center.

    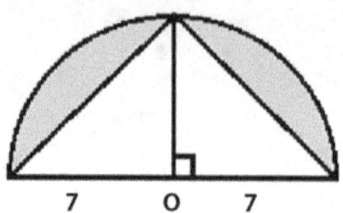

    A. 14.0

    B. 28.0

    C. 55.9

    D. 104.9

    E. 153.9

14. A garden measures 25 m by 40 m, including a circular fishpond with radius 3 m. What is the area of the garden not including the fishpond?

    A. 101.7 m^2

    B. 111.2 m^2

    C. 971.7 m^2

    D. 981.2 m^2

    E. 990.6 m^2

15. The base of cone A has 3 times as great an area as the base of cone B, but the height of cone A is only $\frac{1}{3}$ the height of cone B. Which statement is true?

   A. Cone A has 9 times the volume of cone B.
   B. Cone A has 3 times the volume of cone B.
   C. Cone A and cone B have the same volume.
   D. Cone B has 3 times the volume of cone A.
   E. Cone B has 9 times the volume of cone A.

16. Find the area of the figure depicted below.

   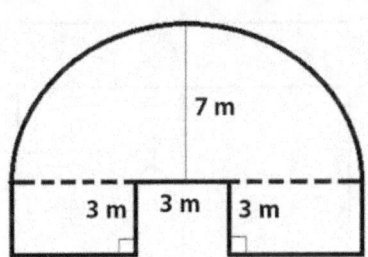

   A. 109.9 m²
   B. 118.9 m²
   C. 142.9 m²
   D. 144.9 m²
   E. 186.9 m²

17. State the domain of the function $f(x) = \dfrac{2x - 14}{x^2 - 9}$.

   A. $x \neq 3$
   B. $x \neq 3, 7$
   C. $x \neq 3, -3$
   D. $x \neq 7$
   E. $x = 3, -3, 7$

18. Which of the following is a factor of the expression $6x^2 - 5x - 14$?

   A. $3x + 7$
   B. $6x + 7$
   C. $6x - 7$
   D. $6x - 5$
   E. $x + 2$

19. Solve for x by factoring:

   $x^2 + x - 6 = 0$

   A. $x = (-3, 2)$
   B. $x = (3, -2)$
   C. $x = (-6, 1)$
   D. $x = (6, -1)$
   E. no real solutions

COLLEGE MATH

20. Which of the following is equivalent to $\sqrt[b]{x^a}$?

    A. $x^{\frac{a}{b}}$

    B. $x^{\frac{b}{a}}$

    C. $a^{\frac{x}{b}}$

    D. $b^{\frac{x}{a}}$

    E. $a^{\frac{b}{x}}$

22. Given $f(x) = 2x+1$ and $g(x) = x^2 - 1$, determine $g(f(x))$.

    A. $4x^2 + 4x - 1$

    B. $4x^2 + 4x + 1$

    C. $4x^2$

    D. $4x^2 - 1$

    E. $4x^2 + 4x$

23. Compute the median for the following data set:
    {9, 11, 18, 13, 12, 21}

    A. 12

    B. 12.5

    C. 13

    D. 14

    E. 15.5

21. Which graph represents the equation?
    $y = x^2 + 3x$?

    A.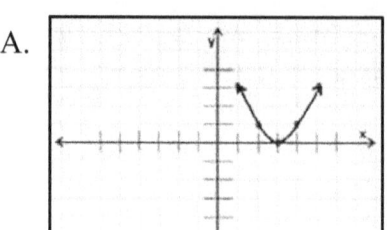

    B.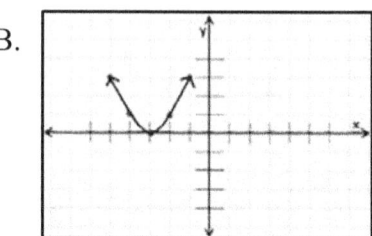

    C.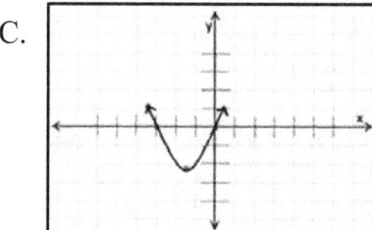

    D.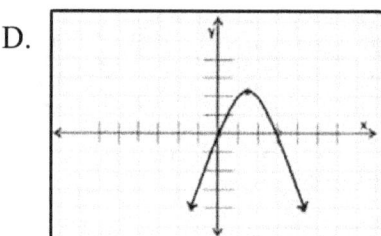

    E.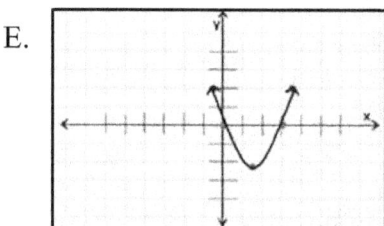

24. What would be the best measure of central tendency for the following collection of high temperatures on 10 successive days?

    {27, 24, 33, 24, 36, 65, 34, 30, 28, 29}

    A. Mean

    B. Either mean or median

    C. Median

    D. Mode

    E. Either median or mode

25. If the correlation between two variables is zero, the association between the two variables is

    A. Negative linear

    B. Positive linear

    C. Quadratic

    D. Direct variation

    E. Random

26. Which of the following is not a valid method of collecting statistical data?

    A. Random sampling

    B. Systematic sampling

    C. Volunteer response

    D. Weighted sampling

    E. Cylindrical sampling

27. A jar contains 3 red marbles and 7 green ones. What is the probability that a marble picked at random from the jar will be red?

    A. $\dfrac{1}{3}$

    B. $\dfrac{1}{7}$

    C. $\dfrac{3}{7}$

    D. $\dfrac{3}{10}$

    E. $\dfrac{7}{10}$

28. A die is rolled several times. What is the probability that a 6 will not appear before the fourth roll of the die?

    A. $\dfrac{125}{216}$

    B. $\dfrac{625}{1296}$

    C. $\dfrac{1}{2}$

    D. $\dfrac{5}{6}$

    E. $\dfrac{1}{216}$

29. There is a 30% chance of rain this Saturday and a 30% chance of rain on Sunday as well. What is the chance of rain on both days?

   A. 9%

   B. 30%

   C. 49%

   D. 60%

   E. 70%

30. Which equation matches the data in the table?

x	3	4	5	6
y	7	8	9	10

   A. $y = 2x - 1$

   B. $y = 2x + 1$

   C. $y = -x + 10$

   D. $y = x + 4$

   E. $y = x - 4$

31. Which table could be generated by the equation? $y = x^2 + 2x - 1$?

   A.
x	1	2	3	4
y	2	5	8	11

   B.
x	1	2	3	4
y	4	9	16	25

   C.
x	1	2	3	4
y	1	5	11	19

   D.
x	1	2	3	4
y	2	7	13	21

   E.
x	1	2	3	4
y	2	7	14	23

32. The fees charged by a parking garage are as follows:

Hours	1	2	3	4	5
Fee	$12	$19	$26	$33	$40

   How would you summarize the fees charged?

   A. $12 an hour

   B. $5 plus $7 per hour

   C. $15 an hour with a $3 discount

   D. $4 plus $8 per hour

   E. $3 plus $9 per hour

33. Which of the following is a solution to $x^2 + 4x + 4 = 25$?

    A. 2

    B. −2

    C. −7

    D. −3

    E. 5

34. Solve the following system of equations:

    $$2x + y = 8$$
    $$4x + 2y = 20$$

    A. $x = 2, y = 4$

    B. $x = 3, y = 1$

    C. $x = 4, y = 0$

    D. no solutions

    E. an infinite number of solutions

35. If an initial deposit of $10,000 is made to a savings account with interest compounded continuously at an annual rate of 6%, how much money is in the account after 5 years?

    A. $13,498.59

    B. $3498.59

    C. $13,382.26

    D. $3,382.26

    E. $13,000.00

36. A dance team comes prepared with a tango, a waltz, a disco number, a salsa routine, and a ballet selection. In how many different orders can they present their routines?

    A. 5

    B. 25

    C. 120

    D. 625

    E. 3125

37. You can choose 3 selections from a buffet table with 8 dishes. How many different plates can you choose?

   A. 6

   B. 24

   C. 56

   D. 336

   E. 6561

38. Leah has 4 blouses, 3 skirts, and 6 pairs of shoes. How many different outfits can she dress herself in?

   A. 12

   B. 13

   C. 24

   D. 72

   E. 720

39. Hiroshi surveys his classmates to find what percent of them come to school on the bus, by car, by subway, by bicycle, or on foot. What is the best way to display his results?

   A. A line graph

   B. A box plot

   C. A stem-and-leaf plot

   D. A scatterplot

   E. A circle graph

40. Which equation could be used as a line of best fit for the scatterplot below?

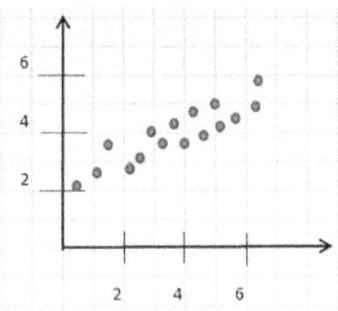

   A. $y = \dfrac{1}{2}x + 2$

   B. $y = 2x + 2$

   C. $y = -2x + 2$

   D. $y = \dfrac{1}{2}x - 2$

   E. $y = \dfrac{1}{2}x + 2$

COLLEGE MATH

41. To find the standard variation of a data set, you first compute the square of the distance of each data item from the mean of all the data items. Then what do you do?

    A. Add all the squared distances and take the square root of the result.

    B. Find the mean of the squared distances and take the square root of the result.

    C. Multiply the squared distances and take the nth root of the result.

    D. Multiply the square root of the sum of the squared distances by the mean of the squared distances.

    E. Multiply the sum of the squared distances by the square root of the mean of the squared distances.

42. In which data set is the mode greater than the median?

    A. {9,11,11,12,14}

    B. {13,15,17,19,21}

    C. {8,11,12,12,19}

    D. {9,9,9,14,20}

    E. {7,11,13,14,14}

43. Of the 200 students in the junior class, 8% are in the Spanish Club. How many juniors are in the Spanish Club?

    A. 4

    B. 8

    C. 16

    D. 20

    E. 25

44. When Olga bought a boat for $1750, she paid an excise tax of $78.75. What was the percent of the tax?

    A. 4.5%

    B. 5.5%

    C. 6.3%

    D. 7%

    E. 7.5%

45. A bank account pays 5% interest yearly. How large an amount would have to be deposited to earn $75 interest in a year?

   A. $375

   B. $875

   C. $1200

   D. $1500

   E. $3750

46. A stock previously trading at $96 a share is now trading at $88 a share. What is the percent of change in the value of the stock?

   A. −8%

   B. −8.3%

   C. −12%

   D. −12.5%

   E. −16%

47. The admission price to tour the Haunted House has been changed from $25 to $30. What is the percent of change in the admission price?

   A. 5%

   B. 16.7%

   C. 20%

   D. 25%

   E. 30%

48. Eileen's Bakery had expenses of $62,500 last year and sales of $68,750. What was the profit as a percent of the expenses?

   A. 6.25%

   B. 10%

   C. 12%

   D. 15%

   E. 16.7%

49. Tim's Typewriters had expenses of $26,200 last year and sales of $19,912. What was the loss as a percent of the expenses?

   A. 7%

   B. 8%

   C. 16.7%

   D. 20%

   E. 24%

50. A stock that had been selling at $30 a share increased its share price by 20%. Later in the day the same stock suffered a 20% decrease in its share price. What was the price at the end of the day?

    A. $24

    B. $28.80

    C. $30

    D. $33

    E. $36

51. A sweater is marked "25% off." The sale price is $36. What was the price before the discount?

    A. $27

    B. $32

    C. $40

    D. $45

    E. $48

52. The sum of $1440 is deposited in a bank which pays 6% simple interest per year. After how many years will there be $1872 in the account?

    A. 2.5 years

    B. 3 years

    C. 4 years

    D. 5 years

    E. 8 years

53. A bank pays 5% interest on deposits, compounded yearly. If $14,000 is deposited, how much will be in the account 3 years later?

    A. $14,350

    B. $15,435

    C. $16,100

    D. $16,206.75

    E. $17,500

54. Which statement is logically equivalent to the following: If it's raining, my roof is leaking.

   A. If my roof isn't leaking, it isn't raining.
   B. If my roof is leaking, it's raining.
   C. If it isn't raining, my roof isn't leaking.
   D. If my roof is leaking, it's not raining
   E. If it's raining, my roof isn't leaking.

55. What is the union of set A and set B?

   Set A: {2,4,5,9,11}
   Set B: {3,5,8,11,13}

   A. {2,3,4,5,5,8,9,11,11,13}
   B. {2,3,4,5,8,9,11,13}
   C. {5,11}
   D. {2,3,4,8,9,13}
   E. {5,9,13,20,24}

56. What is the intersection of set A and set B?

   Set A: {1,3,7,9,10,12,14}
   Set B: {1,4,7,8,11,12,15}

   A. {1,1,3,4,7,7,8,9,10,11,12,12,14,15}
   B. {1,3,4,7,8,9,10,11,12,14,15}
   C. {1,7,12}
   D. {1,1,7,7,12,12}
   E. {3,4,8,9,10,11,14,15}

57. Which statement is NOT implied by the Venn diagram below?

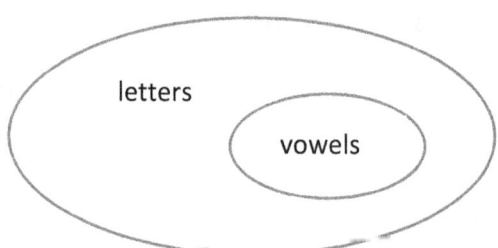

   A. No vowels are not letters.
   B. All vowels are letters.
   C. Some letters are vowels.
   D. Some letters are not vowels.
   E. Some vowels are not letters.

58. A total of 150 students have signed up for musical activities. There are 82 students in the choir and 80 students in the band. How many students are in both the band and the choir?

   A. 12

   B. 24

   C. 42

   D. 70

   E. 162

59. Chris's older brother Mike is 2 years younger than Florence. When Tom's younger sister Rhoda was 8, Chris was 3. Florence is not older than Rhoda. Name the five people in ascending order of age.

   A. Tom, Rhoda, Florence, Mike, Chris

   B. Tom, Florence, Rhoda, Mike, Chris

   C. Chris, Mike, Florence, Rhoda, Tom

   D. Chris, Mike, Rhoda, Florence, Tom

   E. Chris, Rhoda, Mike, Florence, Tom

60. Disprove the following statement by offering a counterexample: "Multiplying two numbers together produces a larger number than either of the two original numbers."

   A. $\sqrt{2} \times \sqrt{2}$

   B. 1.25×1.78

   C. -3×-3

   D. 0.5×0.6

   E. -0.8×-0.3

COLLEGE MATH

ANSWER KEY

Question Number	Correct Answer	Your Answer	Question Number	Correct Answer	Your Answer
1	B		31	E	
2	E		32	B	
3	B		33	C	
4	D		34	D	
5	E		35	A	
6	A		36	C	
7	B		37	C	
8	C		38	D	
9	C		39	E	
10	D		40	A	
11	B		41	B	
12	E		42	E	
13	B		43	C	
14	C		44	A	
15	C		45	D	
16	A		46	B	
17	C		47	C	
18	B		48	B	
19	A		49	E	
20	A		50	B	
21	E		51	E	
22	B		52	D	
23	C		53	D	
24	C		54	A	
25	E		55	B	
26	E		56	C	
27	D		57	E	
28	A		58	A	
29	A		59	C	
30	D		60	D	

COLLEGE MATH

RATIONALES

1. Which of the following is closed under division?

   I. $\left\{\frac{1}{3},1,3\right\}$
   II. $\{-1,1\}$
   III. $\{-1,0,1\}$

   A. I only

   B. II only

   C. III only

   D. I and II

   E. II and III

The answer is B

Set I is not closed under division, because $\frac{1}{3}$ divided by 3 is $\frac{1}{9}$, a number outside the set.

Set III is not closed under division, because it is not possible to divide either −1 or 1 by 0.

2. Which of the following is always composite if x is an odd positive integer and y is an even positive integer greater than 1?

   A. $x+y$

   B. $|x+y|$

   C. $x+2y$

   D. $3x+y$

   E. $3xy$

The answer is E

$3xy$ must be composite, since 3, x, and y are all factors.

18

COLLEGE MATH

3. Find the LCM of 25, 18, and 24.

   A. 1200

   B. 1800

   C. 2400

   D. 3600

   E. 10,800

The answer is B
The LCM must contain 2 factors of 5 to be a multiple of 25. It must contain 2 factors of 3 and a factor of 2 to be a multiple of 18. And it must contain 3 factors of 2 and a factor of 3 to be a multiple of 24. Therefore, the LCM must contain the following factors: $5 \times 5 \times 3 \times 3 \times 2 \times 2 \times 2 = 1800$

4. Solve for x: $|3x| + 6 = 21$

   A. [9,−5]

   B. [−9,5]

   C. [−5,0,5]

   D. [−5,5]

   E. [−9,9]

The answer is D
Write two equations to express the two possibilities:
$$3x + 6 = 21$$
$$-3x + 6 = 21$$
Solving the two equations gives 5 and −5.

COLLEGE MATH

5. Which graph represents the solution set for $x^2 - 5x > -6$?

   A.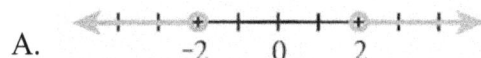

   B.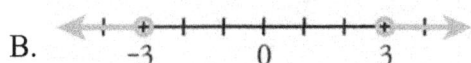

   C.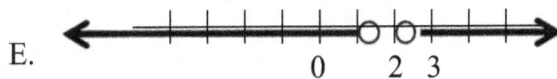

   D.

   E.

The answer is E
Gathering all terms on the left gives $x^2 - 5x + 6 > 0$. Replace the inequality symbol with an equals sign and solve for x: $x = 2$, and $x = 3$. A graph of the parabola makes clear that it is greater than 0 for x-values less than 2 or greater than 3 or greater, but less than 0 when $2 \leq x \leq 3$.

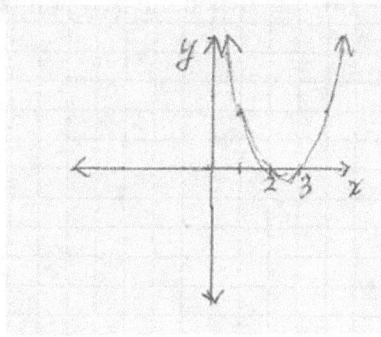

6. What is the equation of the graph shown below?

COLLEGE MATH

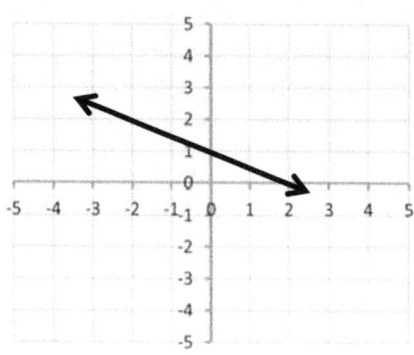

A. $x + 2y = 4$

B. $x - 2y = 4$

C. $2x + y = 4$

D. $x + 2y = -4$

E. $x - 2y = -4$

The answer is A
Replacing *x* with 0 gives a *y*-intercept of 2. Replacing *y* with 0 gives an *x*-intercept of 4. The equation is linear, so a line can be drawn between the two points to complete the graph.

7. Solve the following inequality: $-2x > 4$

   A. $x > -2$

   B. $x < -2$

   C. $x > 2$

   D. $x > -8$

   E. $x < 2$

The answer is B
To solve for *x*, you must divide by –2, but dividing by a negative number reverses the inequality, so the result is $x < -2$.

8. Which equation represents a circle centered on the origin with radius 3?

21

COLLEGE MATH

A. $x^2 + y^2 = 3$

B. $x^2 + y^2 = 6$

C. $x^2 + y^2 = 9$

D. $x^2 + y^2 = 36$

E. $x^2 - y^2 = 9$

The answer is C
The equation for a circle centered on the origin is $x^2 + y^2 = r^2$. Since $r = 3$, the equation in this case is $x^2 + y^2 = 9$.

9. Given that D is a distance, M is a mass, T is a time, and V is a velocity, which of the following units could be used to measure $\frac{MTV}{D}$?

   A. feet

   B. meters

   C. grams

   D. seconds

   E. miles per hour

The answer is C
Try some sample units and see how they interact:
Let the distance be in miles, the mass be in grams, the time be in hours, and the velocity in miles per hour. Then the units to express $\frac{MTV}{D}$ would be $g \times h \times \frac{mi}{h} \times \frac{1}{mi}$ Hours and miles cancel out, leaving only grams.

10. Cubic meters are used to measure which of the following?

COLLEGE MATH

A. Distance

B. Length

C. Area

D. Volume

E. Mass

The answer is D
Distance and length are measured in linear meters. Area is measured in square meters. Mass is not measured in meters of any kind. Of the choices, only volume is measured in cubic meters.

11. **What figure best describes a data set in which many items are clustered near the median value with a smaller number of values less than or greater than the median at greater distances on each side?**

    A. A parabola

    B. A normal curve

    C. A line of best fit

    D. A Cartesian curve

    E. A Newtonian curve

The answer is B
The figure described is a normal curve, called normal because data from the natural world tend to present a shape in which median values are commoner than extreme ones.

12. **If you prove a theorem by showing that an attempt to prove the opposite of the theorem leads to a contradiction, you are using the logical strategy called:**

A. Inductive reasoning

B. Exhaustive proof

C. Proof by attraction

D. Direct proof

E. Indirect proof

The answer is E
Such a proof is called "indirect" because it uses the opposite of the theorem instead of the theorem itself.

13. **Compute the area of the shaded region, given a radius of 7 meters. Point O is the center.**

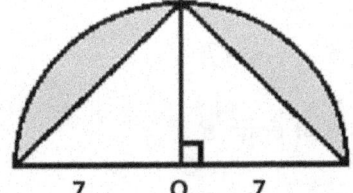

A. 14.0

B. 28.0

C. 55.9

D. 104.9

E. 153.9

The answer is B
The area of the half circle is $\dfrac{49\pi}{2}$ The two triangles are equivalent to a square 7 meters on a side. So the shaded area $= \dfrac{49\pi}{2} - 49 \approx 28.0$.

14. **A garden measures 25 m by 40 m, including a circular fishpond with radius 3 m. What is the area of the garden not including the fishpond?**

A. 101.7 m²

B. 111.2 m²

C. 971.7 m²

D. 981.2 m²

E. 990.6 m²

The answer is C
The area of the garden is $25 \times 40 = 1000$ m². The area of the fishpond is $3^2 \pi \approx 28.3$ m². The difference is about 971.7 m².

15. **The base of cone A has 3 times as great an area as the base of cone B, but the height of cone A is only $\frac{1}{3}$ the height of cone B. Which statement is true?**

    A. Cone A has 9 times the volume of cone B.

    B. Cone A has 3 times the volume of cone B

    C. Cone A and cone B have the same volume.

    D. Cone B has 3 times the volume of cone A.

    E. Cone B has 9 times the volume of cone A.

The answer is C
Let h be the height of cone B and let b be the area of the base of cone B. Using the formula for the volume of a cone, the volume of cone B is $\frac{1}{3}bh$. The base of cone $A = 3b$, while the height of cone $A = \frac{h}{3}$. Therefore, the volume of cone A is $\frac{1}{3}(3b)\left(\frac{h}{3}\right) = \frac{1}{3}bh$, the same as cone B.

16. **Find the area of the figure depicted below.**

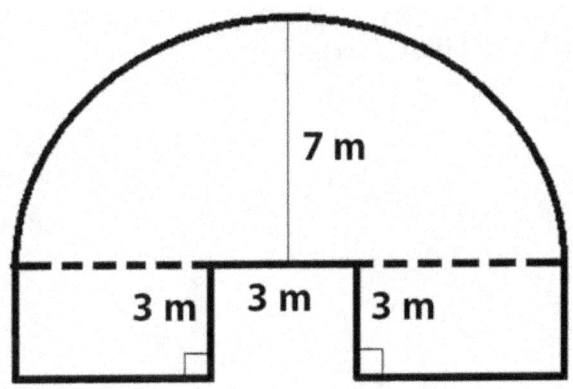

A. 109.9 m²

B. 118.9 m²

C. 142.9 m²

D. 144.9 m²

E. 186.9 m²

The answer is A

The area of the circle is $\dfrac{49\pi}{2}$. The dotted line equals a diameter, twice the length of the radius, or 14. Subtracting the gap of 3 m, the two rectangles add up to a length of 11 m times a width of 3 m So the total area is $\dfrac{49\pi}{2}+33 \approx 109.9$ m².

COLLEGE MATH

17. State the domain of the function $f(x) = \dfrac{2x-14}{x^2-9}$.

   A. $x \neq 3$

   B. $x \neq 3, 7$

   C. $x \neq 3, -3$

   D. $x \neq 7$

   E. $x = 3, -3, 7$

The answer is C
The domain must exclude values of x that would cause the denominator of the function to equal 0. Therefore, both –3 and 3 are excluded from the domain.

18. Which of the following is a factor of the expression $6x^2 - 5x - 14$?

   A. $3x + 7$

   B. $6x + 7$

   C. $6x - 7$

   D. $6x - 5$

   E. $x + 2$

The answer is B
To factor the expression, multiply 6 times 14 to get 84. Then look for two factors of 84 that differ by 5: 7 and 12. Use these factors to rewrite the middle term as $7x - 12x$. You can then factor the expression as $(6x + 7)(x - 2)$.

COLLEGE MATH

19. **Solve for *x* by factoring:** $x^2 + x - 6 = 0$

   A. $x = (-3, 2)$

   B. $x = (3, -2)$

   C. $x = (-6, 1)$

   D. $x = (6, -1)$

   E. no real solutions

The answer is A
Factoring the left side of the equation gives us $(x + 3)(x - 2) = 0$. Setting each factor equal to 0 gives us solutions of –3 and 2.

20. **Which of the following is equivalent to $\sqrt[b]{x^a}$?**

   A. $x^{\frac{a}{b}}$

   B. $x^{\frac{b}{a}}$

   C. $a^{\frac{x}{b}}$

   D. $b^{\frac{x}{a}}$

   E. $a^{\frac{b}{x}}$

The answer is A
Taking the *b*th root of x^a is equivalent to dividing the exponent of x^a by *b*.

COLLEGE MATH

21. **Given** $f(x) = 2x+1$ **and** $g(x) = x^2 - 1$, **determine** $g(f(x))$.

   A. $4x^2 + 4x - 1$

   B. $4x^2 + 4x + 1$

   C. $4x^2$

   D. $4x^2 - 1$

   E. $4x^2 + 4x$

The answer is E
If $f(x) = 2x+1$, $g(f(x)) = (2x+1)^2 - 1 = 4x^2 - 4x.$

22. Compute the median for the following data set: {9, 11, 18, 13, 12, 21}

   A. 12

   B. 12.5

   C. 13

   D. 14

   E. 15.5

The answer is B
In ascending order, the set is {9, 11, 12, 13, 18, 21}
Since there are an even number of data items, the median is halfway between the two most central items when the items are put in ascending order, in this case the third and fourth.

COLLEGE MATH

23. Which graph represents the equation?

$$y = x^2 + 3x?$$

A.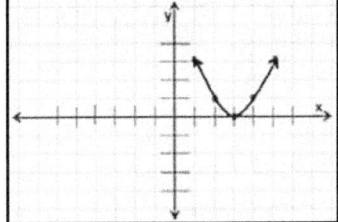

B.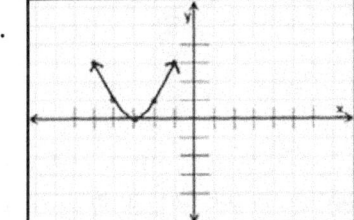

C.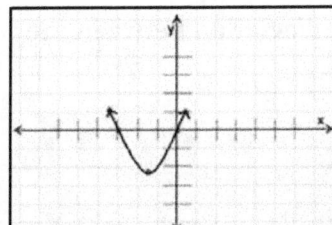

D.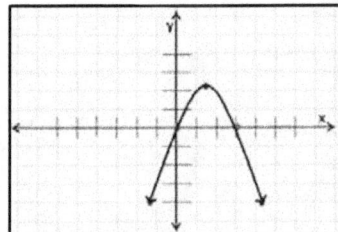

E.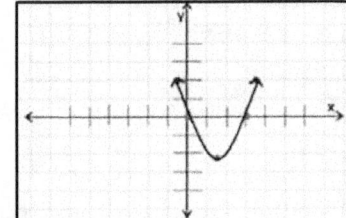

The answer is C

Since $x^2 + 3x$ can be factored as $x(x + 3)$, the function has zeroes at 0 and –3. Since the first term is positive, the parabola opens up. Choice C fits these specifications.

COLLEGE MATH

24. **What would be the best measure of central tendency for the following collection of high temperatures on 10 successive days?**

    {27, 24, 33, 24, 36, 65, 34, 30, 28, 29}

    A. Mean

    B. Either mean or median

    C. Median

    D. Mode

    E. Either median or mode

The answer is C

Since the data contains an outlier, the mean would be skewed too high. The mode is the smallest data item and therefore also not a good representation. The median is the best available representation of the data as a whole.

25. **If the correlation between two variables is zero, the association between the two variables is**

    A. Negative linear

    B. Positive linear

    C. Quadratic

    D. Direct variation

    E. Random

The answer is E

Choices A, B, C, and D all describe some form of correlation between the two variables. Only a random association shows zero correlation.

COLLEGE MATH

26. Which of the following is not a valid method of collecting statistical data?

   A. Random sampling

   B. Systematic sampling

   C. Volunteer response

   D. Weighted sampling

   E. Cylindrical sampling

The answer is E
Choices A, B, C, D describe methods of data collection with varying degrees of potential usefulness and prohibition. There is no such thing as cylindrical sampling.

27. A jar contains 3 red marbles and 7 green ones. What is the probability that a marble picked at random from the jar will be red?

   A. $\dfrac{1}{3}$

   B. $\dfrac{1}{7}$

   C. $\dfrac{3}{7}$

   D. $\dfrac{3}{10}$

   E. $\dfrac{7}{10}$

The answer is D
Three marbles are red out of a total of 10 marbles, yielding a probability of 3/10.

COLLEGE MATH

28. A die is rolled several times. What is the probability that a 6 will not appear before the fourth roll of the die?

A. $\dfrac{125}{216}$

B. $\dfrac{625}{1296}$

C. $\dfrac{1}{2}$

D. $\dfrac{5}{6}$

E. $\dfrac{1}{216}$

The answer is A

Each time the die is rolled, the chance of rolling a number other than 6 is $\dfrac{5}{6}$. The probability that this will happen three times is $\dfrac{5}{6} \times \dfrac{5}{6} \times \dfrac{5}{6} = \dfrac{125}{216}$.

29. There is a 30% chance of rain this Saturday and a 30% chance of rain on Sunday as well. What is the chance of rain on both days?

A. 9%

B. 30%

C. 49%

D. 60%

E. 70%

The answer is A

The probability of two things both happening is the product of the two probabilities: 0.3(0.3) = 0.09 = 9%.

COLLEGE MATH

30. Which equation matches the data in the table?

x	3	4	5	6
y	7	8	9	10

A. $y = 2x - 1$

B. $y = 2x + 1$

C. $y = -x + 10$

D. $y = x + 4$

E. $y = x - 4$

The answer is D
Each *y*-value is 4 greater than the corresponding *x*-value.

31. Which table could be generated by the equation?
$y = x^2 + 2x - 1$?

A.
x	1	2	3	4
y	2	5	8	11

B.
x	1	2	3	4
y	4	9	16	25

C.
x	1	2	3	4
y	1	5	11	19

D.
x	1	2	3	4
y	2	7	14	23

E.
x	1	2	3	4
y	2	7	13	21

The answer is E
Substitute each *x*-value into the equation and see if the result matches the *y*-value. Only in table E do all the *y*-values correspond to the values found by substituting the *x*-values into the equation.

COLLEGE MATH

32. The fees charged by a parking garage are as follows:

Hours	1	2	3	4	5
Fee	$12	$19	$26	$33	$40

How would you summarize the fees charged?

A. $12 an hour

B. $5 plus $7 per hour

C. $15 an hour with a $3 discount

D. $4 plus $8 per hour

E. $3 plus $9 per hour

The answer is B
Each additional hour costs $7 more, so the rate must be $7 an hour, which leaves $5 as the initial fee.

33. Which of the following is a solution to $x^2 + 4x + 4 = 25$?

A. 2

B. −2

C. −7

D. −3

E. 5

The answer is C
Taking the square root of both sides yields $x + 2 = \pm 5$. Therefore, $x = 3$ or −7.

COLLEGE MATH

34. Solve the following system of equations:

$$2x + y = 8$$
$$4x + 2y = 20$$

A. $x = 2, y = 4$

B. $x = 3, y = 1$

C. $x = 4, y = 0$

D. no solutions

E. an infinite number of solutions

The answer is D
Multiply the first equation by 2 and subtract from the second equation. The result is 0 = 4. A system of equations that resolves to an untrue statement has no solutions.

35. If an initial deposit of $10,000 is made to a savings account with interest compounded continuously at an annual rate of 6%, how much money is in the account after 5 years?

A. $13,498.59

B. $3498.59

C. $13,382.26

D. $3,382.26

E. $13,000.00

The answer is A
Continuously compounded interest is calculated using the formula Pe^{rt}, where P is the amount of the principal, r is the annual rate, and t is the time in years.
$10,000 \times e^{0.06 \times 5} = 10,000 e^{0.3} \approx 13,498.59$.

COLLEGE MATH

36. A dance team comes prepared with a tango, a waltz, a disco number, a salsa routine, and a ballet selection. In how many different orders can they present their routines?

    A. 5

    B. 25

    C. 120

    D. 625

    E. 3125

The answer is C
Any of the 5 routines could be the first number. The second number could be any of the remaining 4, the third could be any of the remaining 3, and so on. The total number of choices is $5 \times 4 \times 3 \times 2 \times 1 = 120$.

37. You can choose 3 selections from a buffet table with 8 dishes. How many different plates can you choose?

    A. 6

    B. 24

    C. 56

    D. 336

    E. 6561

The answer is C
Since the order of items on your plate does not matter, it is combinations rather than permutations we need to find. The number of combinations of k items out of n possible selections is given by the formula $\frac{n!}{(n-k)!k!}$. $\frac{8!}{5!3!} = 56$

38. Leah has 4 blouses, 3 skirts, and 6 pairs of shoes. How many different outfits can she dress herself in?

    A. 12

    B. 13

    C. 24

    D. 72

    E. 720

The answer is D
By the Fundamental Counting Principle, the number of different outfits is $4 \times 3 \times 6 = 72$.

39. Hiroshi surveys his classmates to find what percent of them come to school on the bus, by car, by subway, by bicycle, or on foot. What is the best way to display his results?

    A. A line graph

    B. A box plot

    C. A stem-and-leaf plot

    D. A scatterplot

    E. A circle graph

The answer is E
A circle graph is the best way to display what portion of the whole data set is occupied by each item.

COLLEGE MATH

40. Which equation could be used as a line of best fit for the scatterplot below?

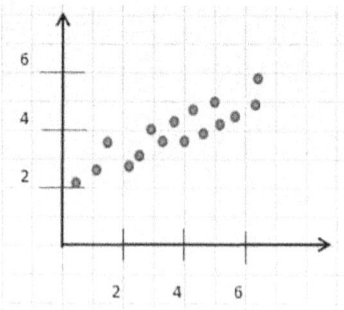

A. $y = \dfrac{1}{2}x + 2$

B. $y = 2x + 2$

C. $y = -2x + 2$

D. $y = \dfrac{1}{2}x - 2$

E. $y = 2x - 2$

The answer is A

The data appear to start around 2, and the y-values are generally rising less fast than the x-values, so the slope appears to be less than 1. $\dfrac{1}{2}x + 2$ is the best fit among the choices.

COLLEGE MATH

41. To find the standard variation of a data set, you first compute the square of the distance of each data item from the mean of all the data items. Then what do you do?

    A. Add all the squared distances and take the square root of the result.

    B. Find the mean of the squared distances and take the square root of the result.

    C. Multiply the squared distances and take the nth root of the result.

    D. Multiply the square root of the sum of the squared distances by the mean of the squared distances.

    E. Multiply the sum of the squared distances by the square root of the mean of the squared distances.

The answer is B
Choice B correctly completes the process of finding a standard variation.

42. In which data set is the mode greater than the median?

    A. {9,11,11,12,14}

    B. {13,15,17,19,21}

    C. {8,11,12,12,19}

    D. {9,9,9,14,20}

    E. {7,11,13,14,14}

The answer is E
In choice E, the median is 13 and the mode 14.

COLLEGE MATH

43. Of the 200 students in the junior class, 8% are in the Spanish Club. How many juniors are in the Spanish Club?

   A. 4

   B. 8

   C. 16

   D. 20

   E. 25

The answer is C
The number of juniors in the Spanish club is 8 % of 200 or 16.

44. When Olga bought a boat for $1750, she paid an excise tax of $78.75. What was the percent of the tax?

   A. 4.5%

   B. 5.5%

   C. 6.3%

   D. 7%

   E. 7.5%

The answer is A
To find the percent of tax, divide the tax by the sales price and multiply by 100.
$\frac{78.75}{1750} \times 100 = 4.5$, so the tax rate is 4.5%.

COLLEGE MATH

45. A bank account pays 5% interest yearly. How large an amount would have to be deposited to earn $75 interest in a year?

   A. $375

   B. $875

   C. $1200

   D. $1500

   E. $3750

The answer is D
If a principal of x dollars earns $75 at 5% interest, then $0.5x = 75$. Multiplying both sides by 20 yields $x = 1500$, so the amount of the principal must be $1500.

46. A stock previously trading at $96 a share is now trading at $88 a share. What is the percent of change in the value of the stock?

   A. −8%

   B. −8.3%

   C. −12%

   D. −12.5%

   E. −16%

The answer is B
The percent of change is found by dividing the amount of the change by the original value, then multiplying by 100. The change is −$8, and the original amount is $96. $\frac{-8}{96} \times 100 \approx -8.3$, So the stock price has changed about −8.3%.

47. The admission price to tour the Haunted House has been changed from $25 to $30. What is the percent of change in the admission price?

    A. 5%

    B. 16.7%

    C. 20%

    D. 25%

    E. 30%

The answer is C

The amount of change is +$5, and the original value is $25. $\frac{5}{25} = \frac{1}{5} = 20\%$.

48. Eileen's Bakery had expenses of $62,500 last year and sales of $68,750. What was the profit as a percent of the expenses?

    A. 6.25%

    B. 10%

    C. 12%

    D. 15%

    E. 16.7%

The answer is B

The amount of change is $6,250. $\frac{6250}{62500} = \frac{1}{10} = 10\%$.

COLLEGE MATH

49. Tim's Typewriters had expenses of $26,200 last year and sales of $19,912. What was the loss as a percent of the expenses?

   A. 7%

   B. 8%

   C. 16.7%

   D. 20%

   E. 24%

The answer is E
The amount of loss was $6288, and 6288./26200 = 24%.

50. A stock that had been selling at $30 a share increased its share price by 20%. Later in the day the same stock suffered a 20% decrease in its share price. What was the price at the end of the day?

   A. $24

   B. $28.80

   C. $30

   D. $33

   E. $36

The answer is B
After the $30 price increased by 20%, it was $36.

COLLEGE MATH

51. A sweater is marked "25% off." The sale price is $36. What was the price before the discount?

    A. $27

    B. $32

    C. $40

    D. $45

    E. $48

The answer is E
If the original price has been decreased by 25%, the sale price is 75% of the original. Solving 36 = 0.75x yields x = 48.

52. The sum of $1440 is deposited in a bank which pays 6% simple interest per year. After how many years will there be $1872 in the account?

    A. 2.5 years

    B. 3 years

    C. 4 years

    D. 5 years

    E. 8 years

The answer is D
After each year is completed, the amount in the account is increased by 0.06(1440) = $86.40 dollars. The number of years required to bring the account to $1872 is $\frac{1872-1440}{86.40} = \frac{432}{86.40} = 5$

53. A bank pays 5% interest on deposits, compounded yearly. If $14,000 is deposited, how much will be in the account 3 years later?

    A. $14,350

    B. $15,435

    C. $16,100

    D. $16,206.75

    E. $17,500

The answer is D

The amount in the account after 3 years will be $14,000 \times 1.05^3 = \$16,206.75$.

54. Which statement is logically equivalent to the following: If it's raining, my roof is leaking.

    A. If my roof isn't leaking, it isn't raining.

    B. If my roof is leaking, it's raining.

    C. If it isn't raining, my roof isn't leaking.

    D. If my roof is leaking, it's not raining

    E. If it's raining, my roof isn't leaking.

The answer is A

The contrapositive of a true statement is also true. In this case, rain always makes my roof leak, so the absence of a leak could only be explained by the absence of rain.

COLLEGE MATH

55. **What is the union of set A and set B?**
    Set A: {2,4,5,9,11}
    Set B: {3,5,8,11,13}

    A. {2,3,4,5,5,8,9,11,11,13}

    B. {2,3,4,5,8,9,11,13}

    C. {5,11}

    D. {2,3,4,8,9,13}

    E. {5,9,13,20,24}

The answer is B
The union of the two sets contains every number that is in either set. Numbers that are in both sets are included only once in the union set.

56. **What is the intersection of set A and set B?**
    Set A: {1,3,7,9,10,12,14}
    Set B: {1,4,7,8,11,12,15}

    A. {1,1,3,4,7,7,8,9,10,11,12,12,14,15}

    B. {1,3,4,7,8,9,10,11,12,14,15}

    C. {1,7,12}

    D. {1,1,7,7,12,12}

    E. {3,4,8,9,10,11,14,15}

The answer is C
The intersection of the two sets contains only those numbers that are in both sets.

57. Which statement is NOT implied by the Venn diagram below?

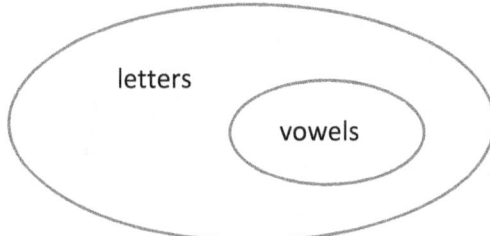

A. No vowels are not letters.

B. All vowels are letters.

C. Some letters are vowels.

D. Some letters are not vowels.

E. Some vowels are not letters.

The answer is E
The diagram shows that the class of vowels is totally included in the class of letters, so there are no vowels that are not letters.

58. A total of 150 students have signed up for musical activities. There are 82 students in the choir and 80 students in the band. How many students are in both the band and the choir?

A. 12

B. 24

C. 42

D. 70

E. 162

The answer is A
The sum of 82 choir members and 80 band members is 162, but only 150 students are involved. The explanation is that 12 students are in both the band and choir, and are therefore counted twice when the two memberships are added.

COLLEGE MATH

59. Chris's older brother Mike is 2 years younger than Florence. When Tom's younger sister Rhoda was 8, Chris was 3. Florence is younger than Rhoda. Name the five people in ascending order of age.

   A. Tom, Rhoda, Florence, Mike, Chris

   B. Tom, Florence, Rhoda, Mike, Chris

   C. Chris, Mike, Florence, Rhoda, Tom

   D. Chris, Mike, Rhoda, Florence, Tom

   E. Chris, Rhoda, Mike, Florence, Tom

The answer is C
Statement 1 allows as to put Chris, Mike, and Florence in ascending order. The second statement allows us to put Rhoda and Tom in ascending order. Since Florence is younger than Rhoda, both Rhoda and Tom are older than Chris, Mike, and Florence, allowing us to put all five in ascending order.

60. Disprove the following statement by offering a counterexample:
    "Multiplying two numbers together produces a larger number than either of the two original numbers."

   A. $\sqrt{2} \times \sqrt{2}$

   B. 1.25×1.78

   C. -3×-3

   D. 0.5×0.6

   E. -0.8×-0.3

The answer is D
The product of 0.5 and 0.6 is 0.3, which is smaller than either of the two original numbers. The other multiplications produce a product larger than either of the numbers multiplied.

COLLEGE ALGEBRA

Description of the Examination

The College Algebra examination covers material that is usually taught in a one-semester college course in algebra. Nearly half of the test is made up of routine problems requiring basic algebraic skills; the remainder involves solving nonroutine problems in which candidates must demonstrate their understanding of concepts. The test includes questions on basic algebraic operations; linear and quadratic equations, inequalities and graphs; algebraic, exponential and logarithmic functions; and miscellaneous other topics. It is assumed that candidates are familiar with currently taught algebraic vocabulary, symbols and notation. The test places little emphasis on arithmetic calculations. However, an online scientific calculator (nongraphing) will be available during the examination.

The examination contains approximately 60 questions to be answered in 90 minutes. Some of these are pretest questions that will not be scored. Any time candidates spend on tutorials and providing personal information is in addition to the actual testing time.

Knowledge and Skills Required

Questions on the College Algebra examination require candidates to demonstrate the following abilities in the approximate proportions indicated.

- Solving routine, straightforward problems (about 50% of the examination)
- Solving nonroutine problems requiring an understanding of concepts and the applications of skills and concepts (about 50% of the examination)

The subject matter of the College Algebra examination is drawn from the following topics. The percentages next to the main topics indicate the approximate percentage of exam questions on that topic.

25% Algebraic Operations
- Operations with exponents
- Factoring and expanding polynomials
- Operations with algebraic expressions
- Absolute value
- Properties of logarithms

25% Equations and Inequalities
- Linear equations and inequalities
- Quadratic equations and inequalities
- Absolute value equations and inequalities
- Systems of equations and inequalities
- Exponential and logarithmic equations

30% Functions and Their Properties
- Definition, interpretation and representation/modeling (graphical, numerical, symbolic, verbal)
- Domain and range
- Evaluation of functions
- Algebra of functions
- Graphs and their properties (including intercepts, symmetry, transformations)
- Inverse functions

20% Number Systems and Operations
- Real numbers
- Complex numbers
- Sequences and series
- Factorials and Binomial Theorem

COLLEGE ALGEBRA

SAMPLE TEST

DIRECTIONS: Read each item and select the best response.

1. Which of the following is a factor of the expression $9x^2 + 6x - 35$?

   A. $3x - 5$

   B. $3x - 7$

   C. $x + 3$

   D. $x - 2$

   E. $x - 3$

2. Given $f(x) = 3x - 2$ and $g(x) = x^2$, determine $g(f(x))$.

   A. $3x^2 - 2$

   B. $9x^2 + 4$

   C. $9x^2 - 12x + 4$

   D. $3x^3 - 2$

   E. $9x^2 - 36$

3. Solve for x: $18 = 4 + |2x|$

   A. $\{-11, 7\}$

   B. $\{-7, 0, 7\}$

   C. $\{-7, 7\}$

   D. $\{-11, 11\}$

   E. $\{-8, 8\}$

4. Solve for x by factoring: $2x^2 - 3x - 2 = 0$

   A. $x = (-1, 2)$

   B. $x = (0.5, -2)$

   C. $x = (-0.5, 2)$

   D. $x = (1, -2)$

   E. $x = (-2, 2)$

5. Which of the following illustrates an inverse property?

   A. $a + b = a - b$

   B. $a + b = b + a$

   C. $a + 0 = a$

   D. $a + (-a) = 0$

   E. $b - a = 0$

6. The conjugate of $4 + 5i$ is

   A. $-4 + 5i$

   B. $4 - 5i$

   C. $4i + 5$

   D. $4i - 5$

   E. $-4 - 5i$

COLLEGE ALGEBRA

7. **Simplify:** $(6+3i)-(4-2i)$

   A. $2+5i$

   B. $2+i$

   C. $10+5i$

   D. $2-2i$

   E. $10-5i$

8. **Simplify:** $\dfrac{10}{1+3i}$

   A. $-1.25(1-3i)$

   B. $1.25(1+3i)$

   C. $1+3i$

   D. $1-3i$

   E. $10+3i$

9. **Solve** $(2b^3 \cdot b^2)^3$

   A. $3b^9$

   B. $2b^8$

   C. $8b^{15}$

   D. $2b^{18}$

   E. $8b^{18}$

10. **Which of the following is incorrect?**

    A. $(x^2 y^3)^2 = x^4 y^6$

    B. $m^2(2n)^3 = 8m^2 n^3$

    C. $\dfrac{m^3 n^4}{m^2 n^2} = mn^2$

    D. $(x+y^2)^2 = x^2 + y^4$

    E. $(2s^{-4}w^4)(7sw^{-5}) = \dfrac{14}{s^3 w}$

11. **Evaluate** $3^{\frac{1}{2}}\left(9^{\frac{1}{3}}\right)$

    A. $27^{\frac{5}{6}}$

    B. $9^{\frac{7}{12}}$

    C. $3^{\frac{5}{6}}$

    D. $3^{\frac{6}{7}}$

    E. $9^{\frac{12}{7}}$

COLLEGE ALGEBRA

12. Simplify: $\dfrac{4x^0 y^{-2} z^3}{4x}$

    A. $\dfrac{z^3}{y^2}$

    B. $\dfrac{z^3}{y^2 x}$

    C. $\dfrac{z^2}{y^3}$

    D. $\dfrac{z^3}{x^2 y}$

    E. $z^3 y^2$

13. The exponential equation $2^5 = 32$ can be written as:

    A. $\log_2(5) = 32$

    B. $\log_{10}(32) = 5$

    C. $\log_5(32) = 2$

    D. $\log_2(32) = 5$

    E. $\log_5(2) = 32$

14. Which equation corresponds to the logarithmic statement $\log_x k = m$?

    A. $x^m = k$

    B. $k^m = x$

    C. $x^k = m$

    D. $m^x = k$

    E. $k^x = m$

15. Solve for x: $\log_6(x-5) + \log_6 x = 2$

    A. $x = 9$

    B. $x = 2, x = 7$

    C. $x = 6$

    D. $x = -2, x = -7$

    E. $x = -4, x = -9$

16. Solve for the slope m and y-intercept: $3x + 2y = 14$

    A. $m = \dfrac{2}{3}, y = 5$

    B. $m = -\dfrac{3}{2}, y = 7$

    C. $m = \dfrac{3}{2}, y = -7$

    D. $m = -\dfrac{2}{3}, y = -5$

    E. $m = 2, y = 7$

COLLEGE ALGEBRA

17. **Simplify:** $-4(-4x-1)-4(7x+3)$

    A. $-44x+16$

    B. $12x-16$

    C. $44x-16$

    D. $-12x-8$

    E. $-11x+2$

18. **Solve** $-2x<5$.

    A. $x<-\dfrac{5}{2}$

    B. $x>-\dfrac{2}{5}$

    C. $x>-\dfrac{5}{2}$

    D. $x>\dfrac{5}{2}$

    E. $x<\dfrac{5}{2}$

19. **Solve** $10 \leq 3x+4 \leq 19$.

    A. $2 \leq x \leq 5$

    B. $-2 \leq x \leq 5$

    C. $x \leq 5$

    D. $x \geq 2$

    E. $-5 \leq x \leq -2$

20. **Solve for** x: $x^2+10x-24=0$

    A. $(-5,12)$

    B. $(-10,8)$

    C. $(12,2)$

    D. $(10,8)$

    E. $(-12,2)$

21. **Find a quadratic equation with roots of 4 and -9.**

    A. $x^2-5x+36=0$

    B. $x^2+5x-36=0$

    C. $4x^2-9x-5=0$

    D. $x^2+4x-9=0$

    E. $5x^2-9x+4=0$

22. **Solve:** $4800 \leq 200x-2x^2$

    A. $-40 \leq x \leq 40$

    B. $x \leq 40$

    C. $40 \leq x \leq 60$

    D. $x=40$

    E. $x=-40$

COLLEGE ALGEBRA

23. Solve: $|3x+2| = 4x+5$

    A. $x = -3$

    B. $x = -1$

    C. $x = 3$

    D. $x = 1$

    E. $x = 6$

24. Solve: $|3x-5| = \dfrac{1}{2}$

    A. $x = -\dfrac{11}{6}, -\dfrac{3}{2}$

    B. $x = -\dfrac{11}{6}, \dfrac{3}{2}$

    C. $x = \dfrac{11}{6}, -\dfrac{3}{2}$

    D. $x = \dfrac{11}{6}, \dfrac{3}{2}$

    E. $x = 11, \dfrac{3}{2}$

25. Solve: $2|3x+9| < 36$

    A. $x < -9$

    B. $x > 3$

    C. $3 < x < 9$

    D. $-9 < x < -3$

    E. $-9 < x < 3$

26. Solve for x and y:
    $4x + 3y = -1$
    $5x + 4y = 1$

    A. $x = -7, y = 9$

    B. $x = 7, y = -9$

    C. $x = 7, y = 9$

    D. $x = -7, y = -9$

    E. $x = y = 7$

27. Which point is in the solution set for the system of inequalities below?
    $x - 7 > 1$
    $y < 2x - 1$

    A. $(-1, -1)$

    B. $(-2, -1)$

    C. $(0, 1)$

    D. $(0, -2)$

    E. $(1, 1)$

28. Solve: $3^{2x-1} = 27$

    A. $x = 2$

    B. $x = -3$

    C. $x = -2$

    D. $x = 3$

    E. $x = \dfrac{2}{3}$

COLLEGE ALGEBRA

29. Solve: $\log_b(x^2) = \log_b(2x-1)$

    A. $x = -2$

    B. $x = 1$

    C. $x = -1$

    D. $x = 2$

    E. $x = 4$

30. Solve: $\log_2(x) + \log_2(x-2) = 3$

    A. $x = 4$

    B. $x = -4, 2$

    C. $x = -4, -2$

    D. $x = 4, -2$

    E. $x = 2$

31. If $f(x) = -3x + 8$, find $f(5)$.

    A. 23

    B. −23

    C. 7

    D. −7

    E. 21

32. Find the zeros of the function $h(x) = \dfrac{x-9}{x+2}$.

    A. $\{9\}$

    B. $\{-2\}$

    C. $\left\{-\dfrac{9}{2}\right\}$

    D. $\{-2, 9\}$

    E. This function has no zeros.

33. Which number line shows the solution to $7x - 5 \geq 9x - 17$?

    A. ←——+——+——●→
        -6   0   6

    B. ←●——+——●→
        -6   0   6

    C. ←——+——+——○→
        -6   0   6

    D. ←●——+——●→
        -6   0   6

    E. ←●——+——+→
        -6   0   6

34. Which graph represents the equation of $y = x^2 + 3x$?

A.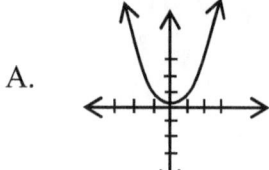

B.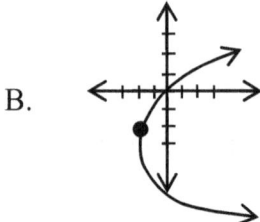

C.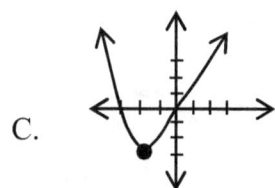

D.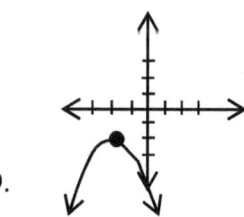

E.

35. Based on the given table, if $y_1 = x^3$, what is the equation for y_2?

x	−2	−1	0	1	2	3
y_1	−8	−1	0	1	8	27
y_2	−18	−11	−10	−9	−2	−17

A. $y_2 = x^5$

B. $y_2 = -x^3$

C. $y_2 = (-x)^3$

D. $y_2 = (x-10)^3$

E. $y_2 = x^3 - 10$

36. Identify the domain and range of the relation:

$$\{(2,-5),(4,31),(11,-11),(-21,3)\}$$

A. Domain is $\{-21\}$, range is $\{-11\}$.

B. Domain is $\{-5,31,-11,3\}$, range is $\{2,4,11,-21\}$.

C. Domain is $\{11\}$, and range is $\{31\}$.

D. Domain and range are indeterminate.

E. Domain is $\{2,4,11,-21\}$, range is $\{-5,31,-11,3\}$.

COLLEGE ALGEBRA

37. Determine the domain of $y=-\sqrt{-2x+3}$.

    A. $x=3$

    B. $x \leq \dfrac{3}{2}$

    C. $x > \dfrac{3}{2}$

    D. $x=2$

    E. $x=0$

38. For the function $h(x)$ whose graph is shown below, select the domain and range.

    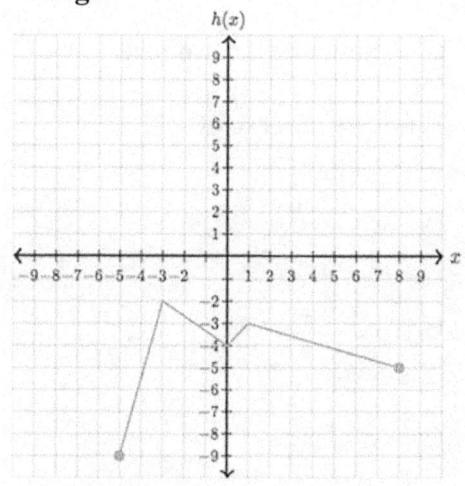

    A. Domain is $-5 \leq x \leq 8$, range is $-9 \leq h \leq -2$.

    B. Domain is -5, range is -5.

    C. Range is $-5 \leq x \leq 8$, domain is $-9 \leq h \leq 2$.

    D. Domain is $x \geq -5$, range is $h \geq -9$.

    E. Domain is 8, range is -2.

39. Given $f(x)=3x^2-7x+5$, find $f(-4)$.

    A. -71

    B. 25

    C. 81

    D. -25

    E. 71

40. For $h(x)=3x^2+ax-1$, $h(3)=8$, find the value of a.

    A. 6

    B. -6

    C. -18

    D. 18

    E. 27

41. Given $f(x)=3x^2-7x+5$, find $\dfrac{f(x+h)-f(x)}{h}$

    A. $7h$

    B. $6xh-7$

    C. $6x+3h-7$

    D. $3x+6h+7$

    E. $5x$

42. Find the x- and y-intercepts for $5x - 3y = 15$.

   A. $x = 0, y = 0$

   B. $x = -3, y = 5$

   C. $x = -1, y = 5$

   D. $x = -5, y = 3$

   E. $x = 3, y = -5$

43. Which of the figures is a reflection of the triangle shown?

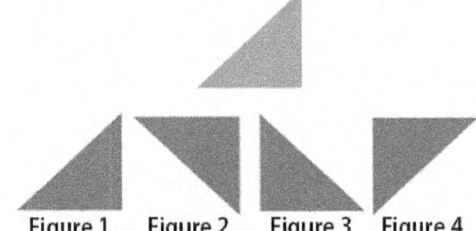

   A. Figure 1 and Figure 4

   B. Figure 4 and Figure 3

   C. Figure 2 and Figure 1

   D. Figures 2, 3 and 4

   E. Figure 1 and Figure 2

44. Name the transformation shown.

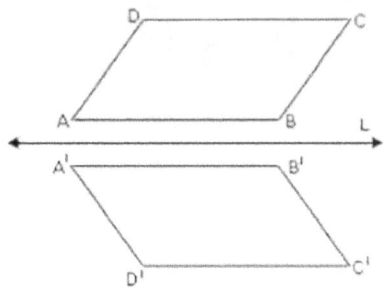

   A. Translation

   B. Rotation

   C. Reflection

   D. Dilation

   E. Cannot be determined

45. Find the inverse of $y = 3x - 2$.

   A. $y = \dfrac{1}{3x - 2}$

   B. $y = \dfrac{x + 2}{3}$

   C. $x = \dfrac{y + 2}{3}$

   D. $y - 3x - 2 = 0$

   E. $3x - y - 2 = 0$

COLLEGE ALGEBRA

46. Find the inverse of $f(x) = -\frac{1}{3}x + 1$

    A. $f^{-1}(x) = 1$

    B. $f^{-1}(x) = 3x$

    C. $f^{-1}(x) = 3x - 3$

    D. $f^{-1}(x) = -3x + 3$

    E. $f^{-1}(x) = x^2$

47. If $f(x) = 3x - 2$ and $g(x) = \frac{x}{3} + \frac{2}{3}$, which of the following is true:

    A. $f(x)$ is the inverse of $g(x)$.

    B. $f(x) = g^{-1}(x)$

    C. There is no connection between $f(x)$ and $g(x)$.

    D. $g(x) = f(x)$

    E. A and B

48. Identify the real numbers in the list: $1.67, \pi, \sqrt{5}, 0$

    A. All

    B. $1.67, \sqrt{5}, 0$

    C. $1.67, 0$

    D. 0

    E. None

49. Which of the following is false?

    A. Every rational number is a real number.

    B. Every imaginary number is a real number.

    C. Every integer is a whole number.

    D. Every integer is a real number.

    E. Every natural number is positive.

50. Which selection below is NOT a real number?

    A. -3

    B. $0.6666...$

    C. $\frac{\pi}{2}$

    D. $3 + \sqrt{2}$

    E. $3i$

51. Simplify $\sqrt{-9}$.

    A. -3

    B. $-3i$

    C. $3i$

    D. 3

    E. 0

COLLEGE ALGEBRA

52. Simplify $(i)(2i)(-3i)$.

   A. $6i$

   B. $-6i^3$

   C. $-6i$

   D. 0

   E. $6i^3$

53. Simplify i^{17}.

   A. $17i$

   B. i

   C. $-17i$

   D. $-i$

   E. 1

54. List the first four terms of the following sequence, beginning with $n=0$.

$$A_n = \frac{(-1)^n}{(n+1)!}$$

   A. $\frac{1}{2}, 1, \frac{3}{2}, 2$

   B. $-1, -\frac{1}{2}, 0, \frac{1}{2}$

   C. $0, 1, 2, 3$

   D. $1, -\frac{1}{2}, \frac{1}{6}, -\frac{1}{24}$

   E. $0, -1, -\frac{1}{2}, \frac{2}{3}$

55. Expand the following series and find the sum:

$$\sum_{n=0}^{4} 2n$$

   A. 20

   B. 8

   C. 16

   D. 4

   E. 32

56. Write the series in sigma notation:
$-3+0+9+24+45+72+105$

   A. $\sum_{a=0}^{6} 3a^2$

   B. $\sum_{a=0}^{6} 3a^2 - 3$

   C. $\sum_{a=0}^{6} a^2 - 3$

   D. $\sum_{a=1}^{6} 3a^2 - 1$

   E. $\sum_{a=0}^{5} a^2 - 3$

COLLEGE ALGEBRA

57. Find $\dfrac{8!}{6!2!}$

   A. $\dfrac{2}{3}$

   B. $\dfrac{4}{6}$

   C. 28

   D. 48

   E. 24

58. Expand the binomial $(2x+3y)^4$

   A. $16x^4+24x^3y+36x^2y^2+54xy^3+81y^4$
      $2x^4+6x^3y+6x^2y^2+6xy^3+3y^4$

   B. $16x^4+81y^4$

   C. $16x^4+96x^3y+216x^2y^2+216xy^3+81y^4$

   D. $16x^4+24x^3y^3+36x^2y^2+54xy+81y^4$

   E. $x^4+4x^3y+6x^2y^2+4xy^3+y^4$

59. Evaluate the determinant of the matrix:
   $$\begin{pmatrix} -2 & 4 \\ -4 & 3 \end{pmatrix}$$

   A. 10

   B. −24

   C. 4

   D. −10

   E. 24

60. Evaluate the determinant of the matrix for $y=4$. $\begin{pmatrix} -5y & 3y \\ y-1 & y-3 \end{pmatrix}$

   A. 35

   B. 12

   C. −56

   D. −12

   E. 56

COLLEGE ALGEBRA

ANSWER KEY

Question Number	Correct Answer	Your Answer
1	A	
2	C	
3	C	
4	C	
5	D	
6	B	
7	A	
8	D	
9	C	
10	D	
11	C	
12	B	
13	D	
14	A	
15	A	
16	B	
17	D	
18	C	
19	A	
20	E	
21	B	
22	C	
23	B	
24	D	
25	E	
26	A	
27	D	
28	A	
29	B	
30	A	

Question Number	Correct Answer	Your Answer
31	D	
32	A	
33	B	
34	C	
35	E	
36	E	
37	B	
38	A	
39	C	
40	B	
41	C	
42	E	
43	D	
44	C	
45	B	
46	D	
47	E	
48	A	
49	E	
50	E	
51	C	
52	A	
53	B	
54	D	
55	A	
56	B	
57	C	
58	C	
59	A	
60	C	

COLLEGE ALGEBRA

RATIONALES

1. Which of the following is a factor of the expression $9x^2+6x-35$?

   A. $3x-5$

   B. $3x-7$

   C. $x+3$

   D. $x-2$

   E. $x-3$

The answer is A
The trinomial can be factored into two binomials, one with addition and one containing subtraction. The factors of 9 to use are 3 and 3 and 7 and 5 are used for 35.
$(3x-5)(3x+7)$ checks when multiplying back through: $9x^2 + 21x - 15x - 35 = 9x^2 + 6x - 35$

2. Given $f(x)=3x-2$ and $g(x)=x^2$, determine $g(f(x))$.

   A. $3x^2-2$

   B. $9x^2+4$

   C. $9x^2-12x+4$

   D. $3x^3-2$

   E. $9x^2-36$

The answer is C
Evaluate: $g(f(x)) = g(3x-2) = (3x-2)^2$
Simplify by expanding: $(3x-2)(3x-2)$
$9x^2 - 6x - 6x + 4$ which simplifies to choice C

COLLEGE ALGEBRA

3. **Solve for** x: $18 = 4 + |2x|$

   A. $\{-11, 7\}$

   B. $\{-7, 0, 7\}$

   C. $\{-7, 7\}$

   D. $\{-11, 11\}$

   E. $\{-8, 8\}$

The answer is C
First isolate the absolute value: $\quad 18 = 4 + |2x|$
$$14 = |2x|$$
Then use the definition of absolute value to set up and solve two equations:
$$2x = 14 \text{ or } 2x = -14$$
$$x = 7, x = -7$$

4. **Solve for** x **by factoring:** $2x^2 - 3x - 2 = 0$

   A. $x = (-1, 2)$

   B. $x = (0.5, -2)$

   C. $x = (-0.5, 2)$

   D. $x = (1, -2)$

   E. $x = (-2, 2)$

The answer is C
Factor the trinomial into one binomial sum and one binomial difference:
$$(2x + 1)(x - 2)$$
Then set each factor equal to zero and solve for x:
$$2x + 1 = 0 \text{ or } x - 2 = 0$$
$$x = -\frac{1}{2}, 2$$

COLLEGE ALGEBRA

5. Which of the following illustrates an inverse property?

   A. $a+b=a-b$

   B. $a+b=b+a$

   C. $a+0=a$

   D. $a+(-a)=0$

   E. $b-a=0$

The answer is D
Choice D represents the sum of a number and its opposite, or additive inverse. This illustrates the inverse property.

6. The conjugate of $4+5i$ is

   A. $-4+5i$

   B. $4-5i$

   C. $4i+5$

   D. $4i-5$

   E. $-4-5i$

The answer is B
For any complex number $a + bi$, the conjugate is defined as $a - bi$.

COLLEGE ALGEBRA

7. **Simplify:** $(6+3i)-(4-2i)$

    A. $2+5i$

    B. $2+i$

    C. $10+5i$

    D. $2-2i$

    E. $10-5i$

The answer is A
To add complex numbers, add the real parts together and the imaginary parts together.
$$(6 + 3i) + -(4 - 2i) = 6 + (-4) + 3i + 2i = 2 + 5i$$

8. **Simplify:** $\dfrac{10}{1+3i}$

    A. $-1.25(1-3i)$

    B. $1.25(1+3i)$

    C. $1+3i$

    D. $1-3i$

    E. $10+3i$

The answer is D
A rational expression with an imaginary denominator must be simplified using the conjugate of the complex denominator: $\dfrac{10}{1+3i} \cdot \dfrac{1-3i}{1-3i} = \dfrac{10-30i}{1-9i^2} = \dfrac{10-30i}{1+9} = \dfrac{10-30i}{10} = 1 - 3i$

COLLEGE ALGEBRA

9. Solve $(2b^3 \cdot b^2)^3$

   A. $3b^9$

   B. $2b^8$

   C. $8b^{15}$

   D. $2b^{18}$

   E. $8b^{18}$

The answer is C
First simplify inside the parenthesis by adding exponents:
$$(2b^3 \cdot b^2)^3 = (2b^5)^3$$
Then raise to the third power, multiplying exponents:
$$(2b^5)^3 = 2^3 b^{15} = 8b^{15}$$

10. Which of the following is incorrect?

    A. $(x^2 y^3)^2 = x^4 y^6$

    B. $m^2(2n)^3 = 8m^2 n^3$

    C. $\dfrac{m^3 n^4}{m^2 n^2} = mn^2$

    D. $(x + y^2)^2 = x^2 + y^4$

    E. $(2s^{-4} w^4)(7sw^{-5}) = \dfrac{14}{s^3 w}$

The answer is D
A power can distribute to a monomial, as seen in choices A and E, but not to a binomial. To find the correct The answer is to D, expand and multiply:
$$(x + y^2)^2 = (x + y^2)(x + y^2) = x^2 + xy^2 + xy^2 + y^4$$

COLLEGE ALGEBRA

11. Evaluate $3^{\frac{1}{2}}\left(9^{\frac{1}{3}}\right)$

   A. $27^{\frac{5}{6}}$

   B. $9^{\frac{7}{12}}$

   C. $3^{\frac{5}{6}}$

   D. $3^{\frac{6}{7}}$

   E. $9^{\frac{12}{7}}$

The answer is C

Rewrite the expression with like bases. $3^{\frac{1}{2}}\left(9^{\frac{1}{3}}\right) = 3^{\frac{1}{2}}\left(3^2\right)^{\frac{1}{3}}$

Then use exponent rules to combine the like bases. $3^{\frac{1}{2}}\left(3^2\right)^{\frac{1}{3}} = 3^{\frac{1}{2}}\left(3^{\frac{2}{3}}\right) = 3^{\left(\frac{3}{6}+\frac{4}{6}\right)} = 3^{\frac{5}{6}}$

COLLEGE ALGEBRA

12. Simplify: $\dfrac{4x^0 y^{-2} z^3}{4x}$

   A. $\dfrac{z^3}{y^2}$

   B. $\dfrac{z^3}{y^2 x}$

   C. $\dfrac{z^2}{y^3}$

   D. $\dfrac{z^3}{x^2 y}$

   E. $z^3 y^2$

The answer is B

Initially, 4/4 reduces to 1 and x^0 also equals 1. Then the expression is $\dfrac{y^{-2} z^3}{x} = \dfrac{z^3}{xy^2}$

13. The exponential equation $2^5 = 32$ can be written as:

   A. $\log_2(5) = 32$

   B. $\log_{10}(32) = 5$

   C. $\log_5(32) = 2$

   D. $\log_2(32) = 5$

   E. $\log_5(2) = 32$

The answer is D

Logarithmic and exponential equations share the following relationship:
$$\text{If } (base)^{exponent} = n, \text{ then } \log_{(base)} n = exponent.$$

COLLEGE ALGEBRA

14. Which equation corresponds to the logarithmic statement $\log_x k = m$?

    A. $x^m = k$

    B. $k^m = x$

    C. $x^k = m$

    D. $m^x = k$

    E. $k^x = m$

The answer is A
See explanation for question 13.

15. Solve for x: $\log_6(x-5) + \log_6 x = 2$

    A. $x = 9$

    B. $x = 2, x = 7$

    C. $x = 6$

    D. $x = -2, x = -7$

    E. $x = -4, x = -9$

The answer is A
Use the log rule: $\log_b(a) + \log_b(c) = \log_b(ac)$ to simplify the equation.
$$\log_6(x-5) + \log_6 x = \log_6 x(x-5) = 2$$

Then rewrite the log as an exponential relationship and solve for x.
$$\log_6 x(x-5) = 2$$
$$6^2 = x(x-5)$$
$$36 = x^2 - 5x$$
$$0 = x^2 - 5x - 36$$
$$0 = (x-9)(x+4), \ x = 9 \text{ or } x = -4$$

However, this solution can yield extraneous solutions, so the the answer must be checked.

COLLEGE ALGEBRA

$$\log_6(9-5) + \log_6 9 \overset{?}{=} 2$$
$$\log_6(4)(9) \overset{?}{=} 2 \qquad\qquad \log_6(-4-5) + \log_6(-4) \overset{?}{=} 2$$
$$\log_6 36 = 2$$

The second portion of the check fails, as the log of a negative number is undefined. So the only solution to the problem is x = 9

16. Solve for the slope m and y-intercept: $3x + 2y = 14$

A. $m = \dfrac{2}{3}, y = 5$

B. $m = -\dfrac{3}{2}, y = 7$

C. $m = \dfrac{3}{2}, y = -7$

D. $m = -\dfrac{2}{3}, y = -5$

E. $m = 2, y = 7$

The answer is B

Put the given equation into slope intercept form, y=mx + b, where m is the slope and b the y intercept.

$$3x + 2y = 14$$
$$2y = -3x + 14$$
$$y = -\dfrac{3}{2}x + 7$$

COLLEGE ALGEBRA

17. Simplify: $-4(-4x-1)-4(7x+3)$

   A. $-44x+16$

   B. $12x-16$

   C. $44x-16$

   D. $-12x-8$

   E. $-11x+2$

The answer is D
Use the distributive property to begin simplifying; then collect like terms.
$$-4(-4x-1)-4(7x+3)$$
$$16x+4-28x-12$$
$$-12x-8$$

18. Solve $-2x<5$.

   A. $x<-\dfrac{5}{2}$

   B. $x>-\dfrac{2}{5}$

   C. $x>-\dfrac{5}{2}$

   D. $x>\dfrac{5}{2}$

   E. $x<\dfrac{5}{2}$

The answer is C
To solve the inequality, divide both sides by -2. This step, however, requires a reversal of the inequality symbol, resulting in choice C.

COLLEGE ALGEBRA

19. Solve $10 \leq 3x + 4 \leq 19$.

    A. $2 \leq x \leq 5$

    B. $-2 \leq x \leq 5$

    C. $x \leq 5$

    D. $x \geq 2$

    E. $-5 \leq x \leq -2$

The answer is A
The first equation solving step used to isolate the x is the subtraction of 4. In a conjunction, the subtraction, as well as the division of 3 following, must be performed on all three parts of the inequality.

$$10 \leq 3x + 4 \leq 19$$
$$6 \leq 3x \leq 15$$
$$2 \leq x \leq 5$$

20. Solve for x: $x^2 + 10x - 24 = 0$

    A. $(-5, 12)$

    B. $(-10, 8)$

    C. $(12, 2)$

    D. $(10, 8)$

    E. $(-12, 2)$

The answer is E
Factor the trinomial and set each factor equal to zero to solve for x.
$$x^2 + 10x - 24 = 0$$
$$(x + 12)(x - 2) = 0$$
$$x + 12 = 0 \text{ or } x - 2 = 0$$

COLLEGE ALGEBRA

21. Find a quadratic equation with roots of 4 and −9.

A. $x^2 - 5x + 36 = 0$

B. $x^2 + 5x - 36 = 0$

C. $4x^2 - 9x - 5 = 0$

D. $x^2 + 4x - 9 = 0$

E. $5x^2 - 9x + 4 = 0$

The answer is B
If r is a root of a polynomial, then (x – r) is a factor.
$$(x - 4)(x + 9) = 0$$
$$x^2 - 4x + 9x - 36 = 0$$
$$x^2 + 5x - 36 = 0$$

22. Solve: $4800 \leq 200x - 2x^2$

A. $-40 \leq x \leq 40$

B. $x \leq 40$

C. $40 \leq x \leq 60$

D. $x = 40$

E. $x = -40$

The answer is C
Start the solution process by setting the inequality less than zero. $2x^2 - 200x + 4800 \leq 0$
One approach is to then graph the parabolic function on a calculator, and, after adjusting the window appropriately, find that the parabola is below the x axis, or less than zero, between 40 and 60.
Alternatively, factor and solve the inequality: $2x^2 - 200x + 4800 \leq 0$
$$x^2 - 100x + 2400 \leq 0$$
$$(x - 40)(x - 60) \leq 0$$
But this work indicates that 40 and 60 are boundaries to a solution interval. Values must be tested in order to determine the actual values where the inequality is less than zero.

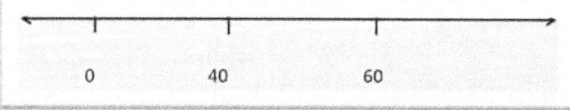

For instance, test a value greater than 60, like 70:

COLLEGE ALGEBRA

$2(70)^2 - 200(70) + 4800 = 600$ which is greater than zero. So the solution does not exist in this interval. Continue with the remaining intervals as shown on the number line to conclude that the polynomial is less than zero between 40 and 60.

23. **Solve:** $|3x+2| = 4x+5$

   A. $x = -3$

   B. $x = -1$

   C. $x = 3$

   D. $x = 1$

   E. $x = 6$

The answer is B
Use the definition of absolute value to set up two equations.

$$3x + 2 = 4x + 5 \text{ or } 3x + 2 = -4x - 5$$
$$-x = 3 \text{ or } -7x = 7$$
$$x = -3 \text{ or } x = -1$$

Check for extraneous solutions:

$\|3x+2\| = 4x+5$	$\|3x+2\| = 4x+5$
$\|3(-3)+2\| \, ? \, 4(-3)+5$	$\|3(-1)+2\| \, ? \, 4(-1)+5$
$\|-9+2\| \, ? -12+5$	$\|-3+2\| \, ? -4+5$
$7 \neq -7$	$1 = 1$

Therefore -3 is not a solution while -1 is.

COLLEGE ALGEBRA

24. **Solve:** $|3x-5| = \dfrac{1}{2}$

   A. $x = -\dfrac{11}{6}, -\dfrac{3}{2}$

   B. $x = -\dfrac{11}{6}, \dfrac{3}{2}$

   C. $x = \dfrac{11}{6}, -\dfrac{3}{2}$

   D. $x = \dfrac{11}{6}, \dfrac{3}{2}$

   E. $x = 11, \dfrac{3}{2}$

The answer is D
Use the definition of absolute value to set up two equations.

$$3x - 5 = \dfrac{1}{2} \qquad\qquad 3x - 5 = -\dfrac{1}{2}$$

$$3x = \dfrac{11}{2} \qquad\qquad 3x = \dfrac{9}{2}$$

$$x = \dfrac{11}{6} \qquad\qquad x = \dfrac{3}{2}$$

COLLEGE ALGEBRA

25. **Solve:** $2|3x+9| < 36$

    A. $x < -9$

    B. $x > 3$

    C. $3 < x < 9$

    D. $-9 < x < -3$

    E. $-9 < x < 3$

The answer is E
First isolate the absolute value, then set up a conjunction to solve.
$$2|3x+9| < 36$$
$$|3x+9| < 18$$
$$-18 < 3x+9 < 18$$
$$-27 < 3x < 9$$
$$-9 < x < 3$$

26. **Solve for x and y:**
    $4x+3y = -1$
    $5x+4y = 1$

    A. $x = -7, y = 9$

    B. $x = 7, y = -9$

    C. $x = 7, y = 9$

    D. $x = -7, y = -9$

    E. $x = y = 7$

The answer is A

Using the elimination method: $4x+3y = -1 \xrightarrow{-4} -16x - 12y = 4$
$5x+4y = 1 \xrightarrow{3} 15x + 12y = 3$

After combining the two new equations, -x = 7 or x = -7. Substitute into one equation to find y. 4(-7) + 3y = -1, y = 9. Therefore the solution to the system is (-7, 9).

COLLEGE ALGEBRA

27. Which point is in the solution set for the system of inequalities below?

$x - 7 < 1$
$y < 2x - 1$

A. $(-1, -1)$

B. $(-2, -1)$

C. $(0, 1)$

D. $(0, -2)$

E. $(1, 1)$

The answer is D
Only point D satisfies both equations algebraically: $0 - 7 < 1$, $-2 < 2(0) - 1$. Additionally, a graph shows that only point D is within the shaded solution region. (Note that point E is on the line, which is not part of the solution region)

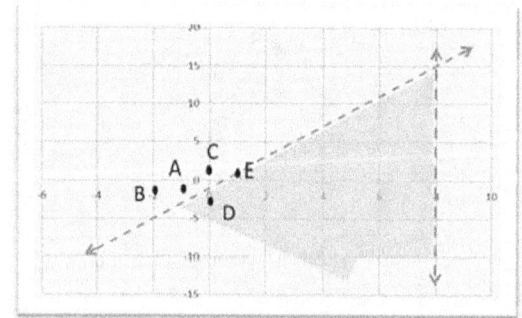

28. Solve: $3^{2x-1} = 27$

A. $x = 2$

B. $x = -3$

C. $x = -2$

D. $x = 3$

E. $x = \dfrac{2}{3}$

The answer is A
Since $27 = 3^3$, the equation can be rewritten: $3^{2x-1} = 3^3$. With the same base on each side, the exponents must be equal. $2x - 1 = 3$, so $x = 2$.

COLLEGE ALGEBRA

29. **Solve:** $\log_b(x^2) = \log_b(2x-1)$

   A. $x = -2$

   B. $x = 1$

   C. $x = -1$

   D. $x = 2$

   E. $x = 4$

The answer is B
Since the log is of the same base on each side of the equation, the arguments are equal.
$$x^2 = 2x - 1$$
$$x^2 - 2x + 1 = 0$$
$$(x-1)(x-1) = 0, x = 1$$

30. **Solve:** $\log_2(x) + \log_2(x-2) = 3$

   A. $x = 4$

   B. $x = -4, 2$

   C. $x = -4, -2$

   D. $x = 4, -2$

   E. $x = 2$

The answer is A
Combine the log expressions into one using the product/sum rule: $\log_b(a) + \log_b(c) = \log_b(ac)$
$$\log_2 x(x-2) = 3$$

Then rewrite the relationship exponentially.
$$2^3 = x(x-2)$$
$$8 = x^2 - 2x$$
$$0 = x^2 - 2x - 8$$
$$0 = (x-4)(x+2), x = 4, -2$$

However, when preparing to check the solutions, x cannot be a negative number as the log function is not defined over negative numbers. Check the positive value for x.

COLLEGE ALGEBRA

$$\log_2(4) + \log_2(4-2) \,?\, 3$$
$$\log_2(4) + \log_2(2) \,?\, 3$$
$$2 + 1 = 3$$

31. **If $f(x) = -3x + 8$, find $f(5)$.**

    A. 23

    B. −23

    C. 7

    D. −7

    E. 21

The answer is D
Evaluate the function for x = 5. $f(5) = -3(5) + 8 = -15 + 8 = -7$

32. **Find the zeros of the function $h(x) = \dfrac{x-9}{x+2}$.**

    A. {9}

    B. {−2}

    C. $\left\{-\dfrac{9}{2}\right\}$

    D. {−2, 9}

    E. This function has no zeros.

The answer is A
The zero of a function is defined as the (x) input required to give the function a (y) value of zero. This function will be zero when the numerator has a value of zero: $x - 9 = 0, x = 9$. When the denominator of this function equals zero, at x = -2, the function will be undefined.

COLLEGE ALGEBRA

33. Which number line shows the solution to $7x - 5 \geq 9x - 17$?

A.
B.
C.
D.
E.

The answer is B

First gather all the x terms on one side of the inequality and the numbers on the other.
$$7x - 5 \geq 9x - 17$$
$$-2x \geq -12$$

When dividing both sides of an inequality by a negative number, the inequality sign is reversed. So division by -2 on both sides results in $x \leq 6$ which is graphed in choice B.

COLLEGE ALGEBRA

34. Which graph represents the equation of $y = x^2 + 3x$?

A.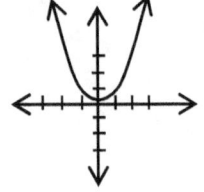

B.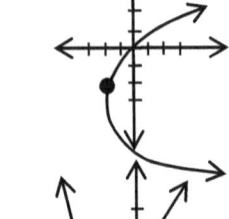

C.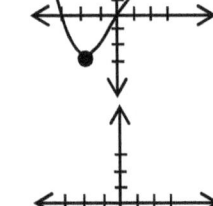

D.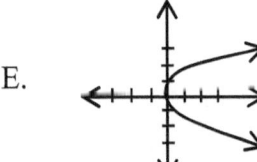

E.

The answer is C

Find the x intercepts of the graph by making y = 0 and solving for x.

$$0 = x^2 + 3x$$

$$0 = x(x+3), x = 0 \text{ or } x = -3$$

Therefore the x intercepts are (0, 0) and (0, -3) which are seen in the graph pictured in choice C.

COLLEGE ALGEBRA

35. Based on the given table, if $y_1 = x^3$, what is the equation for y_2?

x	−2	−1	0	1	2	3
y_1	−8	−1	0	1	8	27
y_2	−18	−11	−10	−9	−2	−17

A. $y_2 = x^5$

B. $y_2 = -x^3$

C. $y_2 = (-x)^3$

D. $y_2 = (x-10)^3$

E. $y_2 = x^3 - 10$

The answer is E
When comparing each y_2 to each y_1, a difference of 10 is observed, resulting in choice E.

36. Identify the domain and range of the relation:

$$\{(2,-5),(4,31),(11,-11),(-21,3)\}$$

A. Domain is $\{-21\}$, range is $\{-11\}$.

B. Domain is $\{-5, 31, -11, 3\}$, range is $\{2, 4, 11, -21\}$.

C. Domain is $\{11\}$, and range is $\{31\}$.

D. Domain and range are indeterminate.

E. Domain is $\{2, 4, 11, -21\}$, range is $\{-5, 31, -11, 3\}$.

The answer is E
In a set of ordered pairs, the domain is made up of the values for x, while the range consists of the y values.

COLLEGE ALGEBRA

37. Determine the domain of $y = -\sqrt{-2x+3}$.

   A. $x = 3$

   B. $x \leq \dfrac{3}{2}$

   C. $x > \dfrac{3}{2}$

   D. $x = 2$

   E. $x = 0$

The answer is B
In order to keep the function defined over the real numbers, the radicand must remain non-negative.
$-2x + 3 \geq 0$
$-2x \geq -3$
$x \leq \frac{3}{2}$ as the inequality is reversed when dividing by a negative number.

COLLEGE ALGEBRA

38. For the function $h(x)$ whose graph is shown below, select the domain and range.

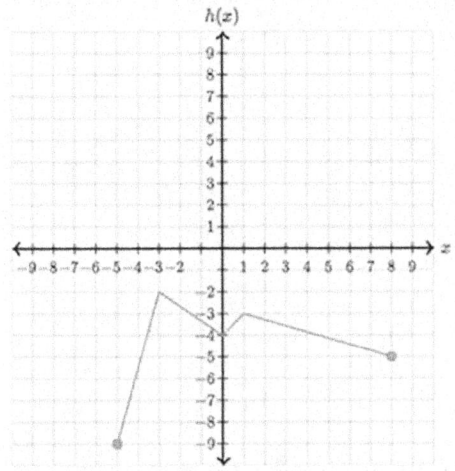

A. Domain is $-5 \leq x \leq 8$, range is $-9 \leq h \leq -2$.

B. Domain is -5, range is -5.

C. Range is $-5 \leq x \leq 8$, domain is $-9 \leq h \leq 2$.

D. Domain is $x \geq -5$, range is $h \geq -9$.

E. Domain is 8, range is -2.

The answer is A
Choice A represents the graphed x values for the domain, and the graphed y values for the range.

39. Given $f(x) = 3x^2 - 7x + 5,$ find $f(-4)$.

A. -71

B. 25

C. 81

D. -25

E. 71

The answer is C
Evaluate the function for x = -4 $f(-4) = 3(-4)^2 - 7(-4) + 5 = 3(16) + 28 + 5 = 81$

COLLEGE ALGEBRA

40. For $h(x) = 3x^2 + ax - 1$, $h(3) = 8$, **find the value of** a.

    A. 6

    B. −6

    C. −18

    D. 18

    E. 27

The answer is B
Evaluate the function for x = 3, then solve for a.
$$h(3) = 8 = 3(3)^2 + a(3) - 1$$
$$8 = 27 + 3a - 1$$
$$-18 = 3a$$
$$a = -6$$

41. Given $f(x) = 3x^2 - 7x + 5$, find $\dfrac{f(x+h) - f(x)}{h}$

    A. $7h$

    B. $6xh - 7$

    C. $6x + 3h - 7$

    D. $3x + 6h + 7$

    E. $5x$

The answer is C

$$\frac{f(x+h) - f(x)}{h} = \frac{3(x+h)^2 - 7(x+h) + 5 - [3x^2 - 7x + 5]}{h}$$
$$= \frac{3x^2 + 6xh + 3h^2 - 7x - 7h + 5 - 3x^2 + 7x - 5}{h}$$
$$= \frac{6xh + 3h^2 - 7h}{h}$$
$$= 6x + 3h - 7$$

COLLEGE ALGEBRA

42. Find the *x*- and *y*- intercepts for $5x - 3y = 15$.

A. $x = 0, y = 0$

B. $x = -3, y = 5$

C. $x = -1, y = 5$

D. $x = -5, y = 3$

E. $x = 3, y = -5$

The answer is E
To find the x intercept, make y = 0: 5x – 3(0) =15, x = 3.
To find the y intercept, make x = 0: 5(0) – 3y = 15, y = -5

43. Which of the figures is a reflection of the triangle shown?

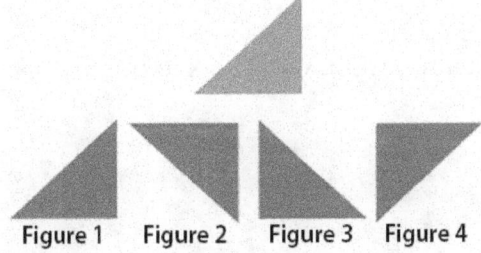

Figure 1   Figure 2   Figure 3   Figure 4

A. Figure 1 and Figure 4

B. Figure 4 and Figure 3

C. Figure 2 and Figure 1

D. Figures 2, 3 and 4

E. Figure 1 and Figure 2

The answer is D
Figure 2 represents a vertical flip, or a reflection over the x axis. Figure 3 represents a horizontal flip, or a reflection over the y axis. Figure 4 is a result of both of these reflections applied.

44. **Name the transformation shown.**

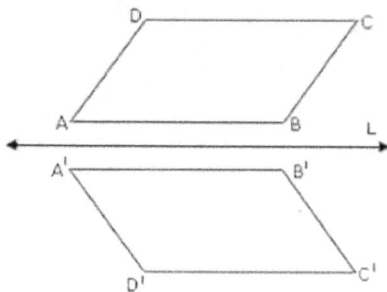

       A. Translation

       B. Rotation

       C. Reflection

       D. Dilation

       E. Cannot be determined

The answer is C
The new image is created from a reflection over the x axis.

45. Find the inverse of $y = 3x - 2$.

A. $y = \dfrac{1}{3x-2}$

B. $y = \dfrac{x+2}{3}$

C. $x = \dfrac{y+2}{3}$

D. $y - 3x - 2 = 0$

E. $3x - y - 2 = 0$

The answer is B
To find the inverse of a function, switch the x and y, then solve for the "new" y.

$x = 3y - 2$
$x + 2 = 3y$
$y = \frac{x+2}{3}$

COLLEGE ALGEBRA

46. **Find the inverse of** $f(x) = -\dfrac{1}{3}x + 1$

    A. $f^{-1}(x) = 1$

    B. $f^{-1}(x) = 3x$

    C. $f^{-1}(x) = 3x - 3$

    D. $f^{-1}(x) = -3x + 3$

    E. $f^{-1}(x) = x^2$

The answer is D
Again, reverse the x and y, where f(x) = y.

$$y = -\frac{1}{3}x + 1$$
$$x = -\frac{1}{3}y + 1$$
$$x - 1 = -\frac{1}{3}y$$
$$-3x + 3 = y$$

Now the "new" y can be noted as $f^{-1}(x)$, which suggests the inverse of $f(x)$.

COLLEGE ALGEBRA

47. If $f(x) = 3x - 2$ and $g(x) = \dfrac{x}{3} + \dfrac{2}{3}$, which of the following is true:

   A. $f(x)$ is the inverse of $g(x)$.

   B. $f(x) = g^{-1}(x)$

   C. There is no connection between $f(x)$ and $g(x)$.

   D. $g(x) = f(x)$

   E. A and B

The answer is E
Choice A and B each communicate that the two functions are inverses of each other. To prove this, derive one from the other, or show that $f(g(x)) = x = g(f(x))$

$$f(g(x)) = 3\left(\dfrac{x}{3} + \dfrac{2}{3}\right) - 2 = x + 2 - 2 = x$$

$$g(f(x)) = \dfrac{3x - 2}{3} + \dfrac{2}{3} = \dfrac{3x}{3} = x$$

48. Identify the real numbers in the list: $1.67, \pi, \sqrt{5}, 0$

   A. All

   B. $1.67, \sqrt{5}, 0$

   C. $1.67, 0$

   D. 0

   E. None

The answer is A
All of the given numbers are real. (None are imaginary)

COLLEGE ALGEBRA

49. Which of the following is false?

  A. Every rational number is a real number.

  B. Every imaginary number is a real number.

  C. Every integer is a whole number.

  D. Every integer is a real number.

  E. Both B and C are false

The answer is E
Choice B is false because no imaginary numbers are real. Choice C is false because some integers are negative, while whole numbers are zero or positive.

50. Which selection below is NOT a real number?

  A. -3

  B. $0.6666...$

  C. $\frac{\pi}{2}$

  D. $3+\sqrt{2}$

  E. $3i$

The answer is E
Choice E shows an imaginary number which is not real.

COLLEGE ALGEBRA

51. Simplify $\sqrt{-9}$.

   A. -3

   B. $-3i$

   C. $3i$

   D. 3

   E. 0

The answer is C
$\sqrt{-9} = \sqrt{9} \cdot \sqrt{-1} = 3i$

52. Simplify $(i)(2i)(-3i)$.

   A. $6i$

   B. $-6i^3$

   C. $-6i$

   D. 0

   E. $6i^3$

The answer is A
$(i)(2i)(-3i) = -6i^3 = -6(-i) = 6i$

53. Simplify i^{17}.

   A. $17i$

   B. i

   C. $-17i$

   D. $-i$

   E. 1

The answer is B
$i^{17} = (i^4)^4 \cdot i = 1i = i$

COLLEGE ALGEBRA

54. List the first four terms of the following sequence, beginning with $n = 0$.

$$A_n = \frac{(-1)^n}{(n+1)!}$$

A. $\frac{1}{2}, 1, \frac{3}{2}, 2$

B. $-1, -\frac{1}{2}, 0, \frac{1}{2}$

C. $0, 1, 2, 3$

D. $1, -\frac{1}{2}, \frac{1}{6}, -\frac{1}{24}$

E. $0, -1, -\frac{1}{2}, \frac{2}{3}$

The answer is D
Evaluate the sequence rule for n = 0, 1, 2, 3

$$\frac{(-1)^0}{(0+1)!}, \frac{(-1)^1}{(1+1)!}, \frac{(-1)^2}{(2+1)!}, \frac{(-1)^3}{(3+1)!}$$

$$\frac{1}{(1)!}, \frac{-1}{(2)!}, \frac{1}{(3)!}, \frac{-1}{(4)!}$$

$$1, \frac{-1}{4}, \frac{1}{6}, \frac{-1}{24}$$

COLLEGE ALGEBRA

55. Expand the following series and find the sum:
$$\sum_{n=0}^{4} 2n$$

A. 20

B. 8

C. 16

D. 4

E. 32

The answer is A
The five terms in the expansion are $0 + 2 + 4 + 6 + 8$ so the sum is 20.

56. Write the series in sigma notation: $-3 + 0 + 9 + 24 + 45 + 72 + 105$

A. $\sum_{a=0}^{6} 3a^2$

B. $\sum_{a=0}^{6} 3a^2 - 3$

C. $\sum_{a=0}^{6} a^2 - 3$

D. $\sum_{a=1}^{6} 3a^2 - 1$

E. $\sum_{a=0}^{5} a^2 - 3$

The answer is B
Working backwards, substituting the start value into the rule for each sum, eliminates all but choice B. That is, choice B is the only rule where a_0 or a_1 matches the first listed term in the sum. Furthermore, evaluating the rule for n = 0 – 6 proves that all 7 terms match confirming that B is the correct choice.

57. Find $\dfrac{8!}{6!2!}$

    A. $\dfrac{2}{3}$

    B. $\dfrac{4}{6}$

    C. 28

    D. 48

    E. 24

The answer is C
Most calculators can handle factorials of this size, however it is practical to know how to simplify an expression of this sort. $\dfrac{8!}{6!2!} = \dfrac{8 \cdot 7 \cdot 6!}{2 \cdot 6!} = \dfrac{8}{2} \cdot 7 = 4 \cdot 7 = 28$

COLLEGE ALGEBRA

58. Expand the binomial $(2x+3y)^4$

A. $16x^4 +24x^3y+36x^2y^2 +54xy^3 +81y^4 \; 2x^4 +6x^3y+6x^2y^2 +6xy^3 +3y^4$

B. $16x^4 +81y^4$

C. $16x^4 +96x^3y+216x^2y^2 +216xy^3 +81y^4$

D. $16x^4 +24x^3y^3 +36x^2y^2 +54xy+81y^4$

E. $x^4 +4x^3y+6x^2y^2 +4xy^3 + y^4$

The answer is C

This problem can be expanded by binomial and trinomial multiplication, using coefficients of Pascal's Triangle, or following the Binomial Expansion Formula.

$$(2x+3y)^4 =(2x+3y)^2 \cdot (2x+3y)^2$$
$$=(4x^2 +12xy+9y^2)(4x^2 +12xy+9y^2)$$
$$=16x^4 +48x^3y+36x^2y^2 +48x^3y+144x^2y^2 +108xy^3 +36x^2y^2 +108xy^3 +81y^4$$
$$=16x^4 +96x^3y+216x^2y^2 +216xy^3 +81y^4$$

Check the third term using the Binomial Expansion Formula:

$$(a+b)^n = \sum_{k=0}^{n} {}_nC_k a^{n-k} b^k$$

To find the third term, k = 2. (since the formula starts with k = 0)
$$(2x+3y)^4: \quad {}_4C_2(2x)^{4-2}(3y)^2 = 6 \cdot 4x^2 \cdot 9y^2 = 216x^2y^2$$

COLLEGE ALGEBRA

59. Evaluate the determinant of the matrix:

$$\begin{pmatrix} -2 & 4 \\ -4 & 3 \end{pmatrix}$$

A. 10

B. −24

C. 4

D. −10

E. 24

The answer is A

Given the matrix $\begin{pmatrix} a & b \\ c & d \end{pmatrix}$, the determinant can be calculated by $ad - cb$.

(-2)(3) − (-4)(4) = -6 − (-16) = -6 + 16 = 10

60. Evaluate the determinant of the matrix for $y=4$. $\begin{pmatrix} -5y & 3y \\ y-1 & y-3 \end{pmatrix}$

A. 35

B. 12

C. −56

D. −12

E. 56

The answer is C

First evaluate the matrix for y = 4. $\begin{pmatrix} -5(4) & 3(4) \\ 4-1 & 4-3 \end{pmatrix} = \begin{pmatrix} -20 & 12 \\ 3 & 1 \end{pmatrix}$

Then use the determinant formula described above: -20(1) − 3(12) = -56

PRECALCULUS

Description of the Examination

The Precalculus examination assesses student mastery of skills and concepts required for success in a first-semester calculus course. A large portion of the exam is devoted to testing a student's understanding of functions and their properties. Many of the questions test a student's knowledge of specific properties of the following types of functions: linear, quadratic, absolute value, square root, polynomial, rational, exponential, logarithmic, trigonometric, inverse trigonometric, and piecewise-defined. Questions on the exam will present these types of functions symbolically, graphically, verbally, or in tabular form. A solid understanding of these types of functions is at the core of all Precalculus courses, and it is a prerequisite for enrolling in calculus and other college-level mathematics courses.

The examination contains approximately 48 questions, in two sections, to be answered in 90 minutes. Any time candidates spend on tutorials and providing personal information is in addition to the actual testing time.
- Section 1: 25 questions, 50 minutes. The use of an online graphing calculator (non-CAS) is allowed for this section. Only some of the questions will require the use of the calculator.
- Section 2: 23 questions, 40 minutes.
- **No calculator is allowed for this section.**

Although most of the questions on the exam are multiple-choice, there are some questions that require students to enter a numerical answer.

Graphing Calculator

A graphing calculator is integrated into the exam software, and it is available to students during Section 1 of the exam. Only some of the questions actually require the graphing calculator. Students are expected to know how and when to make appropriate use of the calculator. The graphing calculator, together with brief video tutorials, is available to students as a free download for a 30-day trial period. Students are expected to download the calculator and become familiar with its functionality prior to taking the exam.
In order to answer some of the questions in the calculator section of the exam, students may be required to use the online graphing calculator in the following ways:
- Perform calculations (e.g., exponents, roots, trigonometric values, logarithms)
- Graph functions and analyze the graphs
- Find zeros of functions
- Find points of intersection of graphs of functions
- Find minima/maxima of functions
- Find numerical solutions to equations
- Generate a table of values for a function

Knowledge and Skills Required

Questions on the examination require candidates to demonstrate the following abilities in the approximate proportions indicated.
- Recalling factual knowledge and/or performing routine mathematical manipulation
- Solving problems that demonstrate comprehension of mathematical ideas and/or concepts
- Solving non-routine problems or problems that require insight, ingenuity, or higher mental processes

The subject matter of the Precalculus examination is drawn from the following topics. The percentages next to the topics indicate the approximate percentage of exam questions on that topic.

PRECALCULUS

20% **Algebraic Expressions, Equations, and Inequalities**
- Ability to perform operations on algebraic expressions
- Ability to solve equations and inequalities, including linear, quadratic, absolute value, polynomial, rational, radical, exponential, logarithmic, and trigonometric
- Ability to solve systems of equations, including linear and nonlinear

15% **Functions: Concept, Properties, and Operations**
- Ability to demonstrate an understanding of the concept of a function, the general properties of functions (e.g., domain, range), function notation, and to perform symbolic operations with functions (e.g., evaluation, inverse functions)

30% **Representations of Functions: Symbolic, Graphical, and Tabular**
- Ability to recognize and perform operations and transformations on functions presented symbolically, graphically, or in tabular form
- Ability to demonstrate an understanding of basic properties of functions and to recognize elementary functions (linear, quadratic, absolute value, square root, polynomial, rational, exponential, logarithmic, trigonometric, inverse trigonometric, and piecewise-defined functions) that are presented symbolically, graphically, or in tabular form

10% **Analytic Geometry**
- Ability to demonstrate an understanding of the analytic geometry of lines, circles, parabolas, ellipses, and hyperbolas

15% **Trigonometry and its Applications***
- Ability to demonstrate an understanding of the basic trigonometric functions and their inverses and to apply the basic trigonometric ratios and identities (in right triangles and on the unit circle)
- Ability to apply trigonometry in various problem-solving contexts

10% **Functions as Models**
- Ability to interpret and construct functions as models and to translate ideas among symbolic, graphical, tabular, and verbal representations of functions

*Note that trigonometry permeates most of the major topics and accounts for more than 15 percent of the exam. The actual proportion of exam questions that requires knowledge of either right triangle trigonometry or the properties of the trigonometric functions is approximately 30-40 percent.

PRECALCULUS

SAMPLE TEST

PART 1: THE FOLLOWING QUESTIONS CAN BE ANSWERED WITH THE AID OF A CALCULATOR

1. Which of the following is not a solution to this system of equations?

$$\begin{cases} y = x^2 \\ y = x + 12 \end{cases}$$

   A. (4, 16)

   B. (-3, 9)

   C. (-3, 4)

   D. (4, -3)

   E. Neither C or D is s solution to the system

2. Solve $\sqrt{n^2 + 16} = 3n$

   A. 2

   B. ±2

   C. ±√2

   D. ±$\frac{4}{3}$

   E. No Real Solution

3. Solve for x such that $0 \leq x \leq 2\pi$.
   $\frac{1}{2}\sin 2x - \frac{\sqrt{2}}{4} = 0$

   A. $\left\{\frac{\pi}{4}\right\}$

   B. $\left\{\frac{\pi}{8}\right\}$

   C. $\left\{\frac{\pi}{8}, \frac{\pi}{4}\right\}$

   D. $\left\{\pm\frac{\pi}{8}, \pm\frac{\pi}{4}\right\}$

   E. $\left\{\frac{\pi}{8}, \frac{3\pi}{8}, \frac{9\pi}{8}, \frac{11\pi}{8}\right\}$

4. Find the solution to the system of equations. $\begin{cases} 4x + 2y = 18 \\ y = -2x + 9 \end{cases}$

   A. No solution

   B. Infinitely many solutions

   C. (2, 1)

   D. (9, 18)

   E. (0, 0)

PRECALCULUS

5. If $h(x) = \frac{3x+4}{x}$ for all real values of $x \neq 0$, find $h^{-1}(x)$

   A. $h^{-1}(x) = \frac{-3x-4}{x}$

   B. $h^{-1}(x) = \frac{x}{3x+4}$

   C. $h^{-1}(x) = \frac{4}{x-3}$

   D. $h^{-1}(x) = \left(\frac{1}{x}\right)(3x+4)$

   E. None of the above represent $h^{-1}(x)$

6. What values of x will keep this function defined over all Real numbers?
   $f(x) = \sqrt{3-2x}$

   A. $\{x \mid x \leq \frac{3}{2}\}$

   B. $\{x \mid x \geq \frac{2}{3}\}$

   C. $\{x \mid -\frac{3}{2} \leq x \leq \frac{3}{2}\}$

   D. $\{x \mid x \neq \frac{3}{2}\}$

   E. All Real numbers, x

7. Which function below does not represent an even function?

   A. $f(x) = 5x^4$

   B. $g(x) = x^2 + 5$

   C. $h(x) = 6x^3$

   D. $q(x) = \cos(x)$

   E. $t(x) = \frac{x^2}{x^4+1}$

8. Based on the given table, if $y_1 = x^3$, what is the equation for y_2?

X	-2	-1	0	1	2	3
y_1	-8	-1	0	1	8	27
y_2	-18	-11	-10	-9	-2	-17

   A. $y_2 = x^5$

   B. $y_2 = -x^3$

   C. $y_2 = (-x)^3$

   D. $y_2 = (x-10)^3$

   E. $y_2 = x^3 - 10$

PRECALCULUS

9. Which choice below makes a true statement regarding the minimum of $f(x) = (x+2)(x-3)(x-12)$?

   A. The minimum value is -2.

   B. The minimum value is equal to zero.

   C. The minimum value is less than zero.

   D. The minimum value occurs between $x = -2$ and $x = 3$.

   E. Choice C and D are both true statements.

10. Find the zeros of the function $h(x) = \frac{x-9}{x+2}$.

    A. $\{9\}$

    B. $\{-2\}$

    C. $\left\{-\frac{9}{2}\right\}$

    D. $\{-2, 9\}$

    E. This function has no zeros.

11. Find the equation of the graph below.

    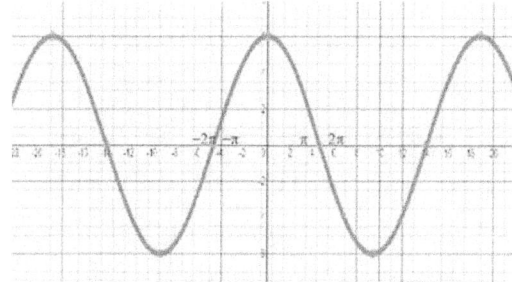

    A. $y = 3\cos(3x)$

    B. $y = 6\cos(6\pi x)$

    C. $y = 6\cos\left(\frac{x}{3}\right)$

    D. $y = 6\sin\left(\frac{x}{3}\right)$

    E. $y = -\sin(3x)$

12. Which of the equations below, when graphed with $y = 10^x$, will show a reflection over the line $= x$?

    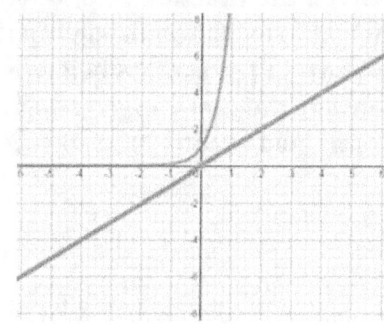

    A. $y = (-10)^x$

    B. $y = -10^x$

    C. $y = 10^{-x}$

    D. $y = \log x$

    E. $y = \ln x$

13. Which interval below represents the domain of the function $t(x) = \sin^{-1} x$?

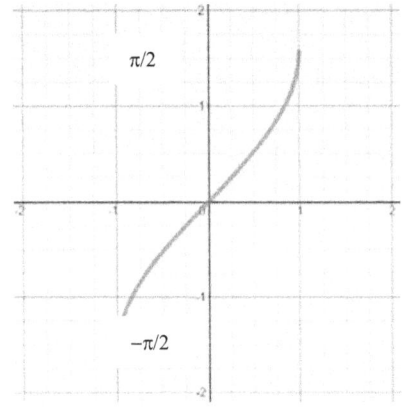

A. [-1, 1]

B. (-1, 1)

C. $[-\pi, \pi]$

D. $(0, 2\pi)$

E. $(-\infty, \infty)$

14. Given the function $f(x) = \sqrt{x} - 10$, which translation to the function, shown in the choices below, would ensure the new range values remained greater than zero?

A. $-f(x)$

B. $f(-x)$

C. $f(x) + 10$

D. $f(x) + 11$

E. None of the above

15. What is true about the following functions: $d(x) = 0.02^x, f(x) = 2.2^x, g(x) = 8.1^x, h(x) = \left(\frac{1}{3}\right)^x$?

A. The graphs of all the functions are increasing.

B. The graphs of all the functions are decreasing.

C. The domain of all the functions is the set of all Real numbers.

D. The range of all the functions is the set of all Real numbers.

E. The graphs each have unique y intercepts.

16. Find the point(s) of intersection of the graphs $f(x) = 4x^2 + 8$ and $g(x) = x^3 + 2x$.

A. (0, 0) and (0, 8)

B. (0, 8)

C. $(2\sqrt{2}, 16)$

D. (4, 72)

E. There is no point of intersection

17. Find the equation of the line that passes through the point (3, 7) and has a slope of $\frac{1}{3}$.

   A. $y = 3x + 7$

   B. $y = \frac{1}{3}x + 7$

   C. $y = \frac{1}{3}x + 6$

   D. $x - 3y + 18 = 0$

   E. Both C and D

18. Find x in the triangle below.

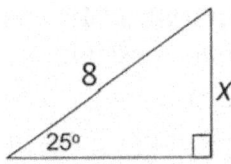

   A. 4

   B. $4\sqrt{2}$

   C. $4\sqrt{3}$

   D. 3.381

   E. 6.928

19. Find $\cos \frac{5\pi}{6}$.

   A. $\frac{1}{2}$

   B. $\frac{\sqrt{2}}{2}$

   C. $\frac{\sqrt{3}}{2}$

   D. $-\frac{1}{2}$

   E. $-\frac{\sqrt{3}}{2}$

20. If a 20 foot ladder needs to make no more than a 65° angle with the ground, how close to the side of the house can the base of the ladder be?

   A. 4 ft.

   B. 4.5 ft.

   C. 8.5 ft.

   D. 9.1 ft.

   E. 10 ft.

PRECALCULUS

21. Which of the following ratios is not equal to $\frac{\sqrt{3}}{3}$?

    A. $\tan 30°$

    B. $\tan 210°$

    C. $\frac{1}{3}\cot 30°$

    D. $\cot 60°$

    E. None of the above (all of the values above equal $\frac{\sqrt{3}}{3}$)

22. A loading dock ramp needs to reach a height 4.5 feet above the ground while making a 10° angle with the ground. How long will the ramp be?

    A. 0.8 ft.

    B. 4.4 ft.

    C. 14.5 ft.

    D. 26 ft.

    E. 45 ft.

23. Which function below is not continuous over the set of real numbers?

    A. $f(x) = |x|$

    B. $g(x) = \sin x$

    C. $h(x) = \frac{1}{x}$

    D. $q(x) = \begin{cases} 0 & \text{for } x > 10 \\ \sqrt{10-x} & \text{for } x \leq 10 \end{cases}$

    E. Both C and D are not continuous

24. Find a polynomial function with zeros at $-3, -\sqrt{2}, 3, \sqrt{2}$.

    A. $f(x) = 3x^4 - 3x^3 + (\sqrt{2})x^2 - (\sqrt{2})x$

    B. $g(x) = x^2(x-3) + x^3(x-\sqrt{2})$

    C. $h(x) = (x^2 + 9)(x^2 + 2)$

    D. $p(x) = x^4 - 11x^2 + 18$

    E. $t(x) = x^4 - 3x^2 + \sqrt{2}$

25. A certain sound wave can be modeled by the sine function. The wave has an amplitude of 10 and a period of 20. Find a possible equation for the wave.

    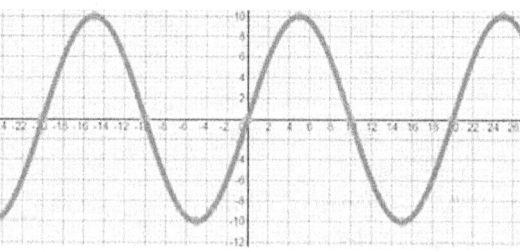

    A. $y = \sin 20x + 10$

    B. $y = \sin(x - 20) + 10$

    C. $y = 10\sin 20x$

    D. $y = 10\sin\frac{x}{20}$

    E. $y = 10\sin\frac{\pi x}{10}$

PRECALCULUS

PART 2: THE FOLLOWING QUESTIONS SHOULD BE ANSWERED WITHOUT THE USE OF A CALCULATOR.

26. Which of the following expressions below is equivalent to $(x-7)^2$?

    A. $x^2 + 49$

    B. $x^2 - 49$

    C. $x^2 - 14x + 49$

    D. $49x^2$

    E. $2x - 14$

27. Solve the equation $\frac{|5a-10|}{3} = \frac{1}{5}$

    A. $\{47, 53\}$

    B. $\left\{\frac{47}{25}, \frac{53}{25}\right\}$

    C. $-\frac{47}{5}$

    D. -47

    E. 53

28. Simplify $\frac{x^2+11x+24}{x+8} + \frac{1}{x}$

    A. $\frac{x^2+3x+1}{x}$

    B. $\frac{x^2+11x+4}{2x+3}$

    C. $\frac{x^2+11x+9}{3x}$

    D. $2x + 4$

    E. 3

29. Select the statement below that explains the best first step to solving the equation $\frac{3x-7}{4} = 5x + 1$.

    A. Add 7 to both sides.

    B. Subtract 3x from both sides.

    C. Divide both sides by 5.

    D. Multiply both sides by $\frac{1}{3}$.

    E. Multiply both sides by 4.

30. Which number line shows the solution to $7x - 5 \geq 9x - 17$?

    A. [number line from -6 to 6, closed circle at 6 extending right]

    B. [number line, shaded from left through 6 with closed circle at 6]

    C. [number line, open circle at 6 extending right]

    D. [number line, shaded between -6 and 6 with closed circles]

    E. [number line, open circle at -6 extending left]

PRECALCULUS

31. Which set of values below represents the range of the function graphed below?

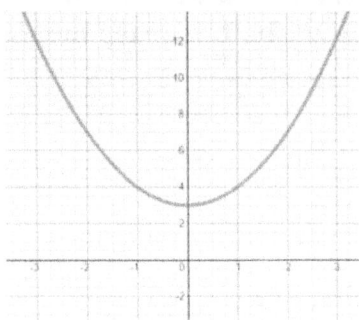

A. All real values of y

B. All real values of y such that $y \geq 3$

C. All real values of y such that $y \geq 0$

D. All real values of y such that $-2 \leq y \leq 2$

E. $\{3, 4, 7, 12, 19, ...\}$

32. Given $g(x) = 2x^2 + 4$, find $g(-3)$.

A. -14

B. -8

C. 16

D. 22

E. 40

33. If $g(x) = x^2 + 9$ and $f(x) = x^2$, find $f(g(x))$.

A. $x + 3$

B. $2x^2 + 9$

C. $x^4 + 9$

D. $x^4 + 81$

E. $x^4 + 18x^2 + 81$

34. Which of the following equations does not represent a function?

A. $y = x^2$

B. $x = y$

C. $x = y - 5$

D. $x = 8$

E. $y = 10$

35. Which function below represents a quadratic equation?

A. $y = x^4$

B. $y = x^3$

C. $y = x^2$

D. $y = x$

E. $y = 0$

110

36. Explain the existence of the dotted lines in the graph of g(x) below.

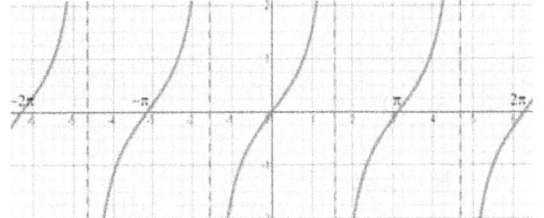

A. g(x) has dotted lines to show it is periodic.

B. g(x) is undefined at the dotted lines.

C. The lines show that g(x) is part linear, part exponential.

D. The lines show that g(x) has a maximum value occurring at $x = \frac{\pi}{2}$.

E. The lines prove that g(x) is not a function.

37. In the given graph, if $y_1 = x^2$, then which is the most likely equation to represent y_2?

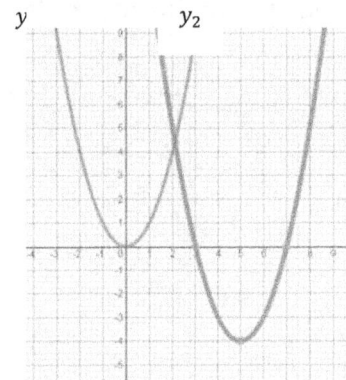

A. $y_2 = 9x^2$

B. $y_2 + 4 = x^2 + 5$

C. $y_2 = (x-5)^2 - 4$

D. $y_2 = (x-5)(x-4)$

E. None of the above

38. Given the piecewise function $f(x) = \begin{cases} x+3 & \text{for } x \geq 0 \\ 5 & \text{for } x < 0 \end{cases}$ find $f(8)$.

A. 2

B. 5

C. 8

D. 11

E. None of the above

39. Which equation below represents an ellipse with center (2, 5), vertical minor axis of length 6, and horizontal major axis of length 10?

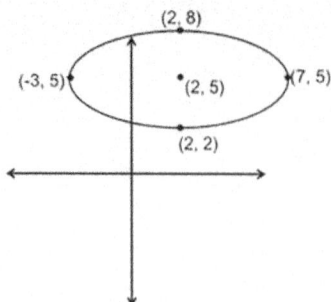

A. $\frac{(x+2)}{10} + \frac{(y+5)}{6} = 1$

B. $\frac{(x-2)^2}{100} + \frac{(y-5)^2}{36} = 1$

C. $\frac{(x-2)^2}{25} + \frac{(y-5)^2}{9} = 1$

D. $(x+3)^2 + (y-7)^2 = 136$

E. $(x+2)^2 + (y+5)^2 = 100$

40. Which equation below represents the positively sloped asymptote for the hyperbola $\frac{x^2}{4} - \frac{y^2}{9} = 1$?

A. $y = 3x + 2$

B. $y = 2x + 3$

C. $y = \frac{2}{3}x$

D. $y = \frac{3}{2}x$

E. $y = \frac{9}{4}x$

41. Find the area of the region bounded by $\begin{cases} x^2 + y^2 = 25 \\ x \geq 0 \\ y \geq 0 \end{cases}$

A. 50π

B. 25π

C. $\frac{25\pi}{4}$

D. $\frac{5\pi}{4}$

E. 25

42. Find h in the triangle pictured below.

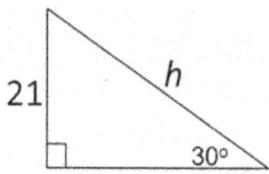

A. 3

B. 7

C. $21\sqrt{2}$

D. $21\sqrt{3}$

E. 42

PRECALCULUS

43. Which equation below represents a parabola with axis of symmetry x = 3?

    A. $y^2 = (x - 3)$

    B. $y = (x - 3)^2$

    C. $x^2 + y^2 = 9$

    D. $y = x^2 + 3$

    E. $y = 3x^2$

44. If x represents the degree measure of an acute angle of a right triangle, and $\cos x = \frac{15}{17}$, find $\tan x$.

    A. $\frac{17}{15}$

    B. $\frac{8}{15}$

    C. $\frac{2}{17}$

    D. 1

    E. 30°

45. Simplify the expression $\frac{2 \sin x (\csc x - \sin x)}{\cot x}$

    A. $\sin 2x$

    B. $2 \sin x$

    C. $(\sin x)^2$

    D. $\tan x$

    E. $1 + \sin x$

46. Simplify the expression: $\sin x \cos x \tan x$.

    A. $\sin x$

    B. $(\sin x)^2$

    C. $\sec x$

    D. $\csc x$

    E. 1

47. Given the function $f(x) = \begin{cases} x^2 + 5 & \text{for } x > 0 \\ x + 5 & \text{for } x \leq 0 \end{cases}$ for what values of x is the graph increasing?

    A. For x > 0

    B. For x < 0

    C. For all Real values of x

    D. For x > 5

    E. The graph is only decreasing

PRECALCULUS

48. Given the following table of values, what are possible equations for y_1 and y_2?

x	y_1	y_2
-3	-3	-3
-2	-2	-2
-1	-1	-1
0	0	0
1	1	-1
2	2	-2
3	3	-3

A. $y_1 = x$, $y_2 = -x$

B. $y_1 = x$, $y_2 = |x|$

C. $y_1 = x$, $y_2 = -|x|$

D. $y_1 = -x$, $y_2 = |x|$

E. $y_1 = |x|$, $y_2 = -|x|$

PRECALCULUS

ANSWER KEY

Question Number	Correct Answer	Your Answer	Question Number	Correct Answer	Your Answer	Question Number	Correct Answer	Your Answer
1	E		17	E		33	E	
2	C		18	D		34	D	
3	E		19	E		35	C	
4	B		20	C		36	B	
5	C		21	E		37	C	
6	A		22	D		38	D	
7	C		23	C		39	C	
8	E		24	D		40	D	
9	C		25	E		41	C	
10	A		26	C		42	E	
11	C		27	B		43	B	
12	D		28	A		44	B	
13	A		29	E		45	A	
14	D		30	B		46	B	
15	C		31	B		47	C	
16	D		32	D		48	C	

PRECALCULUS

RATIONALES

PART 1: THE FOLLOWING QUESTIONS CAN BE ANSWERED WITH THE AID OF A CALCULATOR.

1. Which of the following is not a solution to this system of equations?
$$\begin{cases} y = x^2 \\ y = x + 12 \end{cases}$$

   A. (4, 16)

   B. (-3, 9)

   C. (-3, 4)

   D. (4, -3)

   E. Neither C or D is a solution to the system

The answer is E.
Solve the system by the substitution method, setting $y_1 = y_2$.

$$x^2 = x + 12$$
$$x^2 - x - 12 = 0$$
$$(x - 4)(x + 3) = 0$$
$$x - 4 = 0, x = 4 \quad x + 3 = 0, x = -3$$

Given two solutions for x, substitute them each back into one of the equations to find the y components. This results in the solutions listed in choices A (4, 16) and B (-3, 9). It follows, then, that choices C and D are not solutions.

PRECALCULUS

2. Solve $\sqrt{n^2 + 16} = 3n$

   A. 2

   B. ±2

   C. ±√2

   D. ±$\frac{4}{3}$

   E. No Real Solution

The answer is C.
Square both sides of the equation.
$$n^2 + 16 = 9n^2$$
$$16 = 8n^2$$
$$2 = n^2$$

Take the plus or minus square root of both sides $\pm\sqrt{2} = n$

PRECALCULUS

3. Solve for x such that $0 \leq x \leq 2\pi$. $\frac{1}{2}\sin 2x - \frac{\sqrt{2}}{4} = 0$

   A. $\left\{\frac{\pi}{4}\right\}$

   B. $\left\{\frac{\pi}{8}\right\}$

   C. $\left\{\frac{\pi}{8}, \frac{\pi}{4}\right\}$

   D. $\left\{\pm\frac{\pi}{8}, \pm\frac{\pi}{4}\right\}$

   E. $\left\{\frac{\pi}{8}, \frac{3\pi}{8}, \frac{9\pi}{8}, \frac{11\pi}{8}\right\}$

The answer is E.
Isolate the trigonometric function.

$$\frac{1}{2}\sin 2x - \frac{\sqrt{2}}{4} = 0$$

$$\frac{1}{2}\sin 2x = \frac{\sqrt{2}}{4}$$

$$\sin 2x = \frac{\sqrt{2}}{2}$$

Next list all angles that have a sine value of $\frac{\sqrt{2}}{2}$: $\frac{\pi}{4}, \frac{3\pi}{4}, \frac{9\pi}{4}, \frac{11\pi}{4}, \frac{17\pi}{4}$ ...

Using the given expression for the angle, $2x$

$$2x = \frac{\pi}{4}, \frac{3\pi}{4}, \frac{9\pi}{4}, \frac{11\pi}{4}, \frac{17\pi}{4} \ldots \text{ then } x = \frac{\pi}{8}, \frac{3\pi}{8}, \frac{9\pi}{8}, \frac{11\pi}{8}, \frac{17\pi}{8} \ldots$$

But only the first 4 values listed fall within the given solution specifications, resulting in choice E.

PRECALCULUS

4. **Find the solution to the system of equations.** $\begin{cases} 4x + 2y = 18 \\ y = -2x + 9 \end{cases}$

    A. No solution

    B. Infinitely many solutions

    C. (2, 1)

    D. (9, 18)

    E. (0, 0)

The answer is B.
These two equations represent the same line, as the second can be algebraically manipulated to match the first:

$$4x + 2y = 18$$
$$2y = -4x + 18$$
$$y = -2x + 9$$

Since they are the same line, their intersection, or solution point, is every point on the infinitely long line, making choice B the correct answer.

PRECALCULUS

5. If $h(x) = \frac{3x+4}{x}$ for all real values of $x \neq 0$, find $h^{-1}(x)$

   A. $h^{-1}(x) = \frac{-3x-4}{x}$

   B. $h^{-1}(x) = \frac{x}{3x+4}$

   C. $h^{-1}(x) = \frac{4}{x-3}$

   D. $h^{-1}(x) = \left(\frac{1}{x}\right)(3x+4)$

   E. None of the above represent $h^{-1}(x)$

The answer is C.
Start with $y = \frac{3x+4}{x}$ and replace x and y to find the inverse. $x = \frac{3y+4}{y}$

Then solve for y: $\quad x = \frac{3y+4}{y}$

Cross multiply: $\quad xy = 3y + 4$

Put y terms on same side: $\quad xy - 3y = 4$

Factor out a y: $\quad y(x-3) = 4$

Divide both sides by (x-3) to solve for y: $\quad y = \frac{4}{x-3}$

PRECALCULUS

6. What values of x will keep this function defined over all Real numbers?
   $f(x) = \sqrt{3 - 2x}$

   A. $\{x | x \leq \frac{3}{2}\}$

   B. $\{x | x \geq \frac{2}{3}\}$

   C. $\{x | -\frac{3}{2} \leq x \leq \frac{3}{2}\}$

   D. $\{x | x \neq \frac{3}{2}\}$

   E. All Real numbers, x

The answer is A.
To keep the function defined over the Real Numbers, the radicand must not be negative. Algebraically: $3 - 2x \geq 0$. Solving this inequality yields $x \leq \frac{3}{2}$, or choice A

7. Which function below does not represent an even function?

   A. $f(x) = 5x^4$

   B. $g(x) = x^2 + 5$

   C. $h(x) = 6x^3$

   D. $q(x) = \cos(x)$

   E. $t(x) = \frac{x^2}{x^4+1}$

The answer is C.
According to the definition of an even function, if $f(-x) = f(x)$, then $f(x)$ is an even function. While choice E, for instance, passes this test: $t(-x) = \frac{(-x)^2}{(-x)^4+1} = \frac{x^2}{x^4+1} = t(x)$, choice C fails: $h(-x) = 6(-x)^3 = -6x^3 = -h(x)$

PRECALCULUS

8. Based on the given table, if $y_1 = x^3$, what is the equation for y_2?

X	-2	-1	0	1	2	3
y_1	-8	-1	0	1	8	27
y_2	-18	-11	-10	-9	-2	-17

A. $y_2 = x^5$

B. $y_2 = -x^3$

C. $y_2 = (-x)^3$

D. $y_2 = (x - 10)^3$

E. $y_2 = x^3 - 10$

The answer is E.
When comparing each y_2 to each y_1, a difference of 10 is observed, resulting in choice E.

9. Which choice below makes a true statement regarding the minimum of $f(x) = (x + 2)(x - 3)(x - 12)$?

A. The minimum value is -2.

B. The minimum value is equal to zero.

C. The minimum value is less than zero.

D. The minimum value occurs between $x = -2$ and $x = 3$.

E. Choice C and D are both true statements.

The answer is C.
The lowest portion of the graph occurs between the zeros of 3 and 12 and is substantially less than zero.

PRECALCULUS

10. Find the zeros of the function $h(x) = \frac{x-9}{x+2}$.

   A. $\{9\}$

   B. $\{-2\}$

   C. $\left\{-\frac{9}{2}\right\}$

   D. $\{-2, 9\}$

   E. This function has no zeros.

The answer is A.
The zero of a function is defined as the (x) input required to give the function a (y) value of zero. This function will be zero when the numerator has a value of zero: $x - 9 = 0, x = 9$. When the denominator of this function equals zero, at x = -2, the function will be undefined.

PRECALCULUS

11. Find the equation of the graph below.

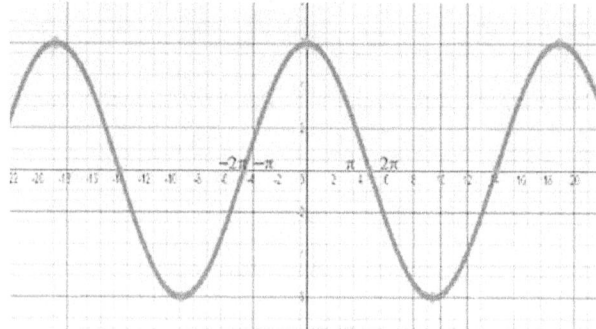

A. $y = 3\cos(3x)$

B. $y = 6\cos(6\pi x)$

C. $y = 6\cos\left(\dfrac{x}{3}\right)$

D. $y = 6\sin\left(\dfrac{x}{3}\right)$

E. $y = -\sin(3x)$

The answer is C.

With a maximum of the graph at x = 0, the image is most likely the graph of the cosine function. Start with the basic formula $y = A\cos(nx)$ where A = the amplitude and n is the period factor. Since the maximum value of the graph is 6, and the graph is balanced over the x axis, the amplitude is 6. The decreasing and increasing portions of the graph represent one period, which is seen to be 6π. Use the following relationship to find the factor, n, needed for the input of the function.

$$\text{Standard period} = n\,(\text{Changed period})$$

$$2\pi = n(6\pi)$$

$$n = \dfrac{2\pi}{6\pi} = \dfrac{1}{3}$$

Therefore the period factor is $\dfrac{1}{3}$ and multiplies with the angle, x. $y = 6\cos\left(\dfrac{1}{3}x\right)$

12. Which of the equations below, when graphed with $y = 10^x$, will show a reflection over the line $= x$?

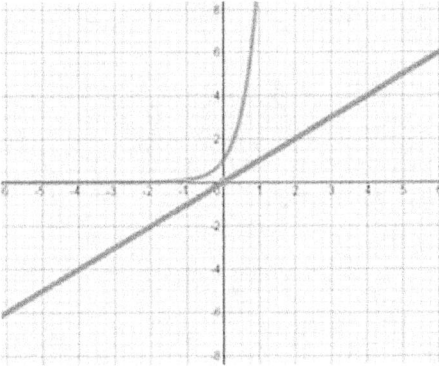

A. $y = (-10)^x$

B. $y = -10^x$

C. $y = 10^{-x}$

D. $y = \log x$

E. $y = \ln x$

The answer is D.
When functions are inverses of each other, their graphs are reflections over the line $y = x$. Since the inverse of $y = 10^x$ is $y = \log x$, the graph of choice D will show a reflection over the line $y = x$ when compared to the original graph.

PRECALCULUS

13. Which interval below represents the domain of the function $t(x) = \sin^{-1} x$?

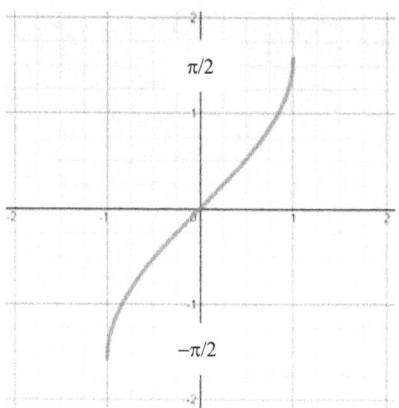

A. $[-1, 1]$

B. $(-1, 1)$

C. $[-\pi, \pi]$

D. $(0, 2\pi)$

E. $(-\infty, \infty)$

The answer is A.

The inverse of the sine function takes an input of a trigonometric value between negative and positive one, and supplies an angle as output. Since the sine of a 90° angle, for instance, is 1, the boundaries of this interval must include 1 and -1, as is shown in choice A. Choice B is the interval infinitely close to, but not including, 1 and -1.

PRECALCULUS

14. Given the function $f(x) = \sqrt{x} - 10$, which translation to the function, shown in the choices below, would ensure the new range values remained greater than zero?

    A. $-f(x)$

    B. $f(-x)$

    C. $f(x) + 10$

    D. $f(x) + 11$

    E. None of the above

The answer is D.
The location of the minimum of $f(x) = \sqrt{x} - 10$ is (0, -10). Choice C would raise this point only up to (0,0). Therefore choice D, which raises the minimum to (0, 1), is the translation that would keep range values positive. Be advised that A is not a valid choice as the x intercept of the original $f(x)$ graph is (100, 0). This point, while far outside the standard viewing range, represents the point where $f(x)$ switches from negative values to positive. Choice A still contains this point. In other words, $-f(x)$ is a graph that does not have a range completely greater than zero.

15. What is true about the following functions:

    $d(x) = 0.02^x, f(x) = 2.2^x, g(x) = 8.1^x, h(x) = \left(\frac{1}{3}\right)^x$?

    A. The graphs of all the functions are increasing.

    B. The graphs of all the functions are decreasing.

    C. The domain of all the functions is the set of all Real numbers.

    D. The range of all the functions is the set of all Real numbers.

    E. The graphs each have unique y intercepts.

The answer is C.
The given functions are all exponential, but those with a base less than one are decreasing graphs, while the ones with a base greater than one are represented by increasing graphs. These exponential functions can be evaluated with any real number, x, but will only yield positive values. Therefore choice C is the best answer.

PRECALCULUS

16. Find the point(s) of intersection of the graphs $f(x) = 4x^2 + 8$ and $g(x) = x^3 + 2x$.

   A. (0, 0) and (0, 8)

   B. (0, 8)

   C. $(2\sqrt{2}, 16)$

   D. (4, 72)

   E. There is no point of intersection

The answer is D.
Algebraically, find the x coordinate of intersection by setting the two function expressions equal to each other.
$$4x^2 + 8 = x^3 + 2x$$

Put all terms on same side equal to zero. $\quad 0 = x^3 + 2x - 4x^2 - 8$
Factor by grouping. $\quad 0 = x(x^2 + 2) - 4(x^2 + 2)$
$\quad 0 = (x - 4)(x^2 + 2)$
Set each factor equal to zero. $\quad x - 4 = 0 \text{ or } x^2 + 2 = 0$

This scenario yields only one real solution for x: = = 4. Evaluate either function to find y: $y = 4(4)^2 + 8 = 72$. Alternatively, use graphing technology to graph the two equations and find their intersection, realizing that the intersection point exists despite the fact that it may appear outside the standard viewing window.

PRECALCULUS

17. Find the equation of the line that passes through the point (3, 7) and has a slope of $\frac{1}{3}$.

    A. $y = 3x + 7$

    B. $y = \frac{1}{3}x + 7$

    C. $y = \frac{1}{3}x + 6$

    D. $x - 3y + 18 = 0$

    E. Both C and D

The answer is E.
First find the equation of the line using point slope form: $y - y_1 = m(x - x_1)$.

$$y - 7 = \frac{1}{3}(x - 3)$$
$$y - 7 = \frac{1}{3}x - 1$$
$$y = \frac{1}{3}x + 6, \text{ choice C}$$

But then know that this equation can be rewritten, first by multiplying through by 3:

$$3y = x + 18$$

then by rearranging terms: $3y - x - 18 = 0$ or $x - 3y + 18 = 0$, choice D.

PRECALCULUS

18. Find *x* in the triangle below.

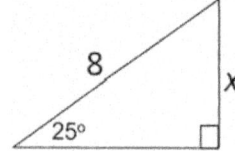

   A. 4

   B. $4\sqrt{2}$

   C. $4\sqrt{3}$

   D. 3.381

   E. 6.928

The answer is D.
The right triangle comes with information representing the sine ratio:
$\sin \theta = \frac{opposite}{hypotenuse}$ or in this case, $\sin 25 = \frac{x}{8}$. Use a calculator to find the decimal approximation for $\sin 25 = .422618 ...$ Multiplication by 8 yields the answer presented in choice D.

19. Find $\cos \frac{5\pi}{6}$.

   A. $\frac{1}{2}$

   B. $\frac{\sqrt{2}}{2}$

   C. $\frac{\sqrt{3}}{2}$

   D. $-\frac{1}{2}$

   E. $-\frac{\sqrt{3}}{2}$

The answer is E.
In terms of the Unit Circle, the angle in question is in the second quadrant, where the cosine is negative. The reference angle at that location is $\frac{\pi}{6}$ or 30° and the $\cos 30 = \frac{\sqrt{3}}{2}$. Combining these two conditions results in choice E.

PRECALCULUS

20. **If a 20 foot ladder needs to make no more than a 65° angle with the ground, how close to the side of the house can the base of the ladder be?**

    A. 4 ft.

    B. 4.5 ft.

    C. 8.5 ft.

    D. 9.1 ft.

    E. 10 ft.

The answer is C.
The problem describes a right triangle, where the ladder is the hypotenuse and the distance from the side of the house is the adjacent side. The corresponding trig ratio equation is $\cos 65 = \frac{x}{20}$ Use a calculator to find $20 \cos 65 \approx 8.5$

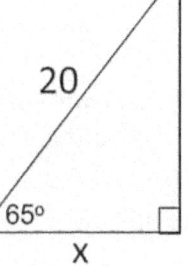

21. **Which of the following ratios is not equal to $\frac{\sqrt{3}}{3}$?**

    A. $\tan 30°$

    B. $\tan 210°$

    C. $\frac{1}{3} \cot 30°$

    D. $\cot 60°$

    E. None of the above (all of the values above equal $\frac{\sqrt{3}}{3}$)

The answer is E.
Unit circle relationships find all the ratios to be equal.

For example: $\sin 30 = \frac{1}{2}, \cos 30 = \frac{\sqrt{3}}{2}$

$$\tan 30 = \frac{\sin 30}{\cos 30} = \frac{\frac{1}{2}}{\frac{\sqrt{3}}{2}} = \frac{1}{\sqrt{3}} = \frac{\sqrt{3}}{3}, \quad \left(\frac{1}{3}\right)\cot 30 = \left(\frac{1}{3}\right)\frac{\cos 30}{\sin 30} = \left(\frac{1}{3}\right)\frac{\frac{\sqrt{3}}{2}}{\frac{1}{2}} = \frac{\sqrt{3}}{3}$$

PRECALCULUS

22. A loading dock ramp needs to reach a height 4.5 feet above the ground while making a 10° angle with the ground. How long will the ramp be?

   A. 0.8 ft.

   B. 4.4 ft.

   C. 14.5 ft.

   D. 26 ft.

   E. 45 ft.

The answer is D.
The situation described can be represented by a right triangle with a height of 4.5 and an unknown hypotenuse.

Set up and solve the ratio: $\sin 10 = \frac{4.5}{x}$

$x \sin 10 = 4.5$

$x = \frac{4.5}{\sin 10} \approx 26$

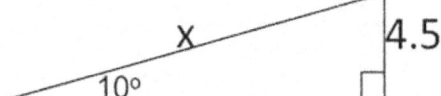

23. Which function below is not continuous over the set of real numbers?

   A. $f(x) = |x|$

   B. $g(x) = \sin x$

   C. $h(x) = \frac{1}{x}$

   D. $q(x) = \begin{cases} 0 & \text{for } x > 10 \\ \sqrt{10-x} & \text{for } x \leq 10 \end{cases}$

   E. Both C and D are not continuous

The answer is C.
In $h(x) = \frac{1}{x}, x \neq 0$ which makes the graph undefined, or discontinuous at x = 0. While the equation listed in choice D is a piecewise function, the fact that the values at the "split" point are equal keeps the function continuous.

PRECALCULUS

24. Find a polynomial function with zeros at -3, $-\sqrt{2}$, 3, $\sqrt{2}$.

   A. $f(x) = 3x^4 - 3x^3 + (\sqrt{2})x^2 - (\sqrt{2})x$

   B. $g(x) = x^2(x-3) + x^3(x-\sqrt{2})$

   C. $h(x) = (x^2 + 9)(x^2 + 2)$

   D. $p(x) = x^4 - 11x^2 + 18$

   E. $t(x) = x^4 - 3x^2 + \sqrt{2}$

The answer is D.
The zero of a function is the same as the root of an equation. If r is a root of a polynomial equation then $(x - r)$ is a factor. Use the 4 given zeros, or roots, to create 4 factors:

$$(x+3)(x-3)(x+\sqrt{2})(x-\sqrt{2})$$

After multiplying the conjugate pairs: $\quad (x^2 - 9)(x^2 - 2)$

After multiplying the binomials: $\quad x^4 - 9x^2 - 2x^2 + 18 \quad$ which simplifies to choice D.

PRECALCULUS

25. A certain sound wave can be modeled by the sine function. The wave has an amplitude of 10 and a period of 20. Find a possible equation for the wave.

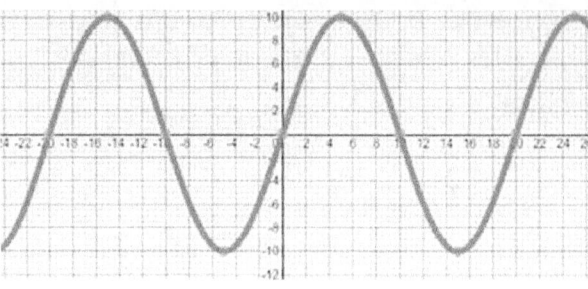

A. $y = \sin 20x + 10$

B. $y = \sin(x - 20) + 10$

C. $y = 10\sin 20x$

D. $y = 10 \sin \frac{x}{20}$

E. $y = 10 \sin \frac{\pi x}{10}$

The answer is E.
Start with the basic formula $y = A\sin(nx)$ where A = the amplitude and n is the period factor. To find *n*, use the relationship: Standard period = *n* (Changed period).

$$2\pi = n(20)$$

$$n = \frac{2\pi}{20} = \frac{\pi}{10}$$

PRECALCULUS

PART 2: THE FOLLOWING QUESTIONS SHOULD BE ANSWERED WITHOUT THE USE OF A CALCULATOR.

26. Which of the following expressions below is equivalent to $(x-7)^2$?

    A. $x^2 + 49$

    B. $x^2 - 49$

    C. $x^2 - 14x + 49$

    D. $49x^2$

    E. $2x - 14$

The answer is C.
The squared binomial can be expanded: $(x-7)(x-7)$
And multiplied: $x^2 - 7x - 7x + 49$
And combined: $x^2 - 14x + 49$

27. Solve the equation $\frac{|5a-10|}{3} = \frac{1}{5}$

    A. $\{47, 53\}$

    B. $\left\{\frac{47}{25}, \frac{53}{25}\right\}$

    C. $-\frac{47}{5}$

    D. -47

    E. 53

The answer is B.
First, isolate the absolute value bars by multiplying both sides of the equation by 3.
$$|5a - 10| = \frac{3}{5}$$

Then set up two equations to represent the definition of absolute value.
$$5a - 10 = \frac{3}{5} \quad \text{and} \quad 5a - 10 = -\frac{3}{5}$$
$$5a = \frac{53}{5} \quad \text{and} \quad 5a = \frac{47}{5}$$
$$a = \frac{53}{25} \quad \text{and} \quad a = \frac{47}{25}$$

28. Simplify $\frac{x^2+11x+24}{x+8} + \frac{1}{x}$

A. $\frac{x^2+3x+1}{x}$

B. $\frac{x^2+11x+4}{2x+3}$

C. $\frac{x^2+11x+9}{3x}$

D. $2x+4$

E. 3

The answer is A.
First factor and cancel in the first portion of the expression.

$$\frac{x^2+11x+24}{x+8} \rightarrow \frac{(x+8)(x+3)}{(x+8)} \rightarrow x+3$$

Then find common denominators to add the two expressions together.

$$(x+3) + \frac{1}{x}$$

$$\frac{x(x+3)}{x} + \frac{1}{x}$$

$$\frac{x^2+3x+1}{x}$$

PRECALCULUS

29. Select the statement below that explains the best first step to solving the equation $\frac{3x-7}{4} = 5x + 1$.

   A. Add 7 to both sides.

   B. Subtract 3x from both sides.

   C. Divide both sides by 5.

   D. Multiply both sides by 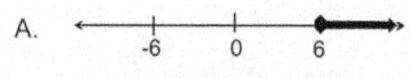 $\frac{1}{3}$.

   E. Multiply both sides by 4.

The answer is E.
While choices A - D do describe steps that may take place later in the equation solving process, they are not advisable as first steps due to the interference of the denominator on the left side of the equation. Choice E as a first step "clears" that denominator so the rest of the steps can occur.

30. Which number line shows the solution to $7x - 5 \geq 9x - 17$?

   A.
   B.
   C.
   D.
   E.

The answer is B.
First gather all the x terms on one side of the inequality and the numbers on the other.

$$7x - 5 \geq 9x - 17$$

$$-2x \geq -12$$

When dividing both sides of an inequality by a negative number, the inequality sign is reversed. So division by -2 on both sides results in $x \leq 6$ which is graphed in choice B.

PRECALCULUS

31. Which set of values below represents the range of the function graphed below?

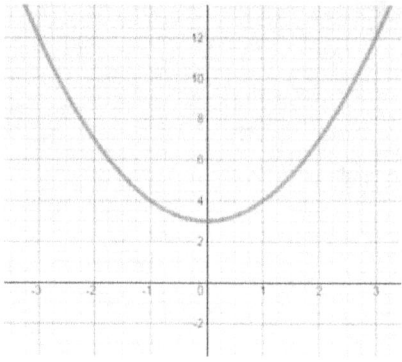

A. All real values of y

B. All real values of y such that $y \geq 3$

C. All real values of y such that $y \geq 0$

D. All real values of y such that $-2 \leq y \leq 2$

E. $\{3, 4, 7, 12, 19, \dots\}$

The answer is B.
While choice E represents the y values of the function created when the function is evaluated for x with the counting numbers, in reality all real numbers can be inserted into the function yielding their corresponding values. There are an infinite number of ordered pairs satisfying this equation, but the y values will never be any smaller than 3, resulting in choice B.

32. Given $g(x) = 2x^2 + 4$, find $g(-3)$.

A. -14

B. -8

C. 16

D. 22

E. 40

The answer is D.
Evaluate: $g(-3) = 2(-3)^2 + 4 = 2(9) + 4 = 18 + 4 = 22$

33. If $g(x) = x^2 + 9$ and $f(x) = x^2$, find $f(g(x))$.

   A. $x + 3$

   B. $2x^2 + 9$

   C. $x^4 + 9$

   D. $x^4 + 81$

   E. $x^4 + 18x^2 + 81$

The answer is E.
Evaluate: $\qquad f(g(x)) = f(x^2 + 9) = (x^2 + 9)^2$

Simplify by expanding:
$$(x^2 + 9)(x^2 + 9)$$
$$x^4 + 9x^2 + 9x^2 + 81$$
$$x^4 + 18x^2 + 81$$

34. Which of the following equations does not represent a function?

   A. $y = x^2$

   B. $x = y$

   C. $x = y - 5$

   D. $x = 8$

   E. $y = 10$

The answer is D.
A function can have only one output, y, for each input, x. A table of ordered pairs for choice D could be

x	8	8	8	8	8
y	-2	0	1	3	6

This shows the (only) input, 8, has multiple outputs, y. Therefore the equation fails to be a function.

PRECALCULUS

35. Which function below represents a quadratic equation?

   A. $y = x^4$

   B. $y = x^3$

   C. $y = x^2$

   D. $y = x$

   E. $y = 0$

The answer is C.
A quadratic equation is defined as being of degree 2.

36. Explain the existence of the dotted lines in the graph of g(x) below.

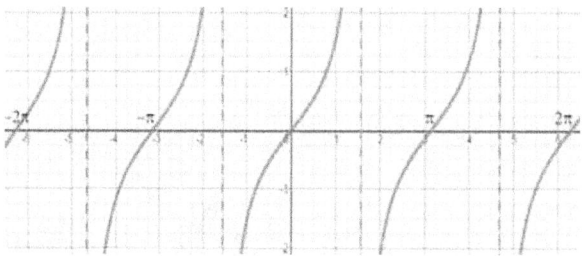

   A. g(x) has dotted lines to show it is periodic.

   B. g(x) is undefined at the dotted lines.

   C. The lines show that g(x) is part linear, part exponential.

   D. The lines show that g(x) has a maximum value occurring at $x = \frac{\pi}{2}$.

   E. The lines prove that g(x) is not a function.

The answer is B.
Dotted lines in a graph represent asymptotes, which indicate places where a graph is trending, but will never reach. In this case, g(x) appears to be the tangent function. The $\tan\left(\frac{\pi}{2}\right)$ is undefined. The curve will stretch infinitely high as it approaches $\frac{\pi}{2}$, but will have no value for that input of the function.

PRECALCULUS

37. In the given graph, if $y_1 = x^2$, then which is the most likely equation to represent y_2?

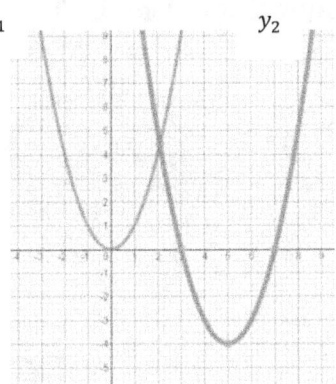

A. $y_2 = 9x^2$

B. $y_2 + 4 = x^2 + 5$

C. $y_2 = (x - 5)^2 - 4$

D. $y_2 = (x - 5)(x - 4)$

E. None of the above

The answer is C.
The vertex of the graph of y_2 is (5, 4). Therefore choice C is the best equation.

38. Given the piecewise function $f(x) = \begin{cases} x + 3 & \text{for } x \geq 0 \\ 5 & \text{for } x < 0 \end{cases}$ find $f(8)$.

A. 2

B. 5

C. 8

D. 11

E. None of the above

The answer is D.
The input value, 8, is greater than zero, so the function is to be evaluated using the first portion of the rule: $f(x) = x + 3$, so $f(8) = 8 + 3 = 11$.

PRECALCULUS

39. Which equation below represents an ellipse with center (2, 5), vertical minor axis of length 6, and horizontal major axis of length 10?

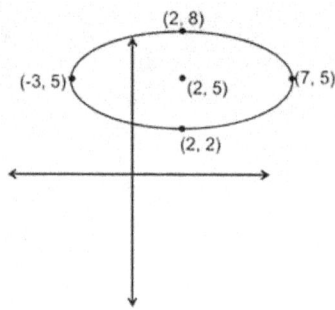

A. $\frac{(x+2)}{10} + \frac{(y+5)}{6} = 1$

B. $\frac{(x-2)^2}{100} + \frac{(y-5)^2}{36} = 1$

C. $\frac{(x-2)^2}{25} + \frac{(y-5)^2}{9} = 1$

D. $(x+3)^2 + (y-7)^2 = 136$

E. $(x+2)^2 + (y+5)^2 = 100$

The answer is C.

The standard form of a horizontal ellipse with center *(h, k)* is $\frac{(x-h)^2}{a^2} + \frac{(y-k)^2}{b^2} = 1$, where a is distance from the center to the vertex on the major axis, and b is the distance from the center to the vertex on the minor axis.

PRECALCULUS

40. Which equation below represents the positively sloped asymptote for the hyperbola $\frac{x^2}{4} - \frac{y^2}{9} = 1$?

A. $y = 3x + 2$

B. $y = 2x + 3$

C. $y = \frac{2}{3}x$

D. $y = \frac{3}{2}x$

E. $y = \frac{9}{4}x$

The answer is D.
The formula for a horizontal hyperbola centered on the origin is $\frac{x^2}{a^2} - \frac{y^2}{b^2} = 1$ where a is the distance from the center to the vertices and b is the distance from the center to the covertices. The asymptotes go through the center of the hyperbola, which in this case is the origin. The slope of each asymptote is $\pm\frac{b}{a}$. Therefore the positive asymptote for this hyperbola, following the linear pattern $y = mx + b$, is $y = \frac{3}{2}x + 0$ or $y = \frac{3}{2}x$.

41. Find the area of the region bounded by $\begin{cases} x^2 + y^2 = 25 \\ x \geq 0 \\ y \geq 0 \end{cases}$

A. 50π

B. 25π

C. $\frac{25\pi}{4}$

D. $\frac{5\pi}{4}$

E. 25

The answer is C.
The given region is the first quadrant portion of a circle, centered on the origin, with radius 5. The full area of such a circle would be $A = \pi r^2 = 25\pi$. The first quadrant section is a quarter of the entire circle, resulting in answer choice C.

PRECALCULUS

42. Find *h* in the triangle pictured below.

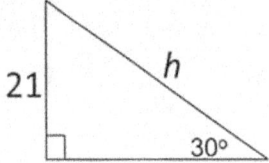

A. 3

B. 7

C. $21\sqrt{2}$

D. $21\sqrt{3}$

E. 42

The answer is E.
In a 30°, 60°, 90° triangle, if the side opposite the 30 is n, then the side opposite the 60 is $n\sqrt{3}$ and the hypotenuse is 2n. Applying that relationship to the given diagram results in the hypotenuse, h, being 2(21) or 42.

43. Which equation below represents a parabola with axis of symmetry x = 3?

A. $y^2 = (x - 3)$

B. $y = (x - 3)^2$

C. $x^2 + y^2 = 9$

D. $y = x^2 + 3$

E. $y = 3x^2$

The answer is B.
Choice B represents a parabola, opening up, with vertex (3, 0). Such a parabola has a line of symmetry at x = 3.

PRECALCULUS

44. If *x* represents the degree measure of an acute angle of a right triangle, and $\cos x = \frac{15}{17}$, find $\tan x$.

    A. $\frac{17}{15}$

    B. $\frac{8}{15}$

    C. $\frac{2}{17}$

    D. 1

    E. 30°

The answer is B.
The ratio can be found without ever knowing the value for the angle, *x*. First, draw a right triangle with a hypotenuse of 17. Arbitrarily choose one of the angles to be *x*, then make the leg adjacent to that side have a length of 15. According to the Pythagorean Theorem, the opposite leg then must be 8. Therefore the tangent ratio, "opposite over adjacent," is $\frac{8}{15}$.

PRECALCULUS

45. Simplify the expression $\dfrac{2\sin x(\csc x - \sin x)}{\cot x}$

A. $\sin 2x$

B. $2\sin x$

C. $(\sin x)^2$

D. $\tan x$

E. $1 + \sin x$

The answer is A.

First rewrite the expression in terms of sine and cosine.

$$\frac{2\sin x\left(\dfrac{1}{\sin x} - \sin x\right)}{\dfrac{\cos x}{\sin x}}$$

Distribute, and change division to multiplication of the reciprocal.

$$2\left(\frac{2\sin x}{\sin x} - (\sin x)^2\right)\left(\frac{\sin x}{\cos x}\right)$$

$$2(1 - (\sin x)^2)\left(\frac{\sin x}{\cos x}\right)$$

Use a Pythagorean Identity.

$$2((\cos x)^2)\left(\frac{\sin x}{\cos x}\right)$$

Divide out the cosine function and apply the double angle identity. $2\sin x \cos x \rightarrow \sin 2x$

PRECALCULUS

46. Simplify the expression: $\sin x \cos x \tan x$.

   A. $\sin x$

   B. $(\sin x)^2$

   C. $\sec x$

   D. $\csc x$

   E. 1

The answer is B.
$\frac{\sin x}{1} \times \frac{\cos x}{1} \times \frac{\sin x}{\cos x} = (\sin x)^2$

47. Given the function $f(x) = \begin{cases} x^2 + 5 & \text{for } x > 0 \\ x + 5 & \text{for } x \leq 0 \end{cases}$ for what values of x is the graph increasing?

   A. For x > 0

   B. For x < 0

   C. For all Real values of x

   D. For x > 5

   E. The graph is only decreasing

The answer is C.
When looking at the graph from left to right, it is first a line with a positive slope, making an increasing graph. The graph is continuous at x = 0, since the point (0, 5) is common to both equations. Then, as the graph continues to the right of the y axis, it is the right half of a parabola, which is also an increasing graph.

PRECALCULUS

48. Given the following table of values, what are possible equations for y_1 and y_2?

x	y_1	y_2
-3	-3	-3
-2	-2	-2
-1	-1	-1
0	0	0
1	1	-1
2	2	-2
3	3	-3

A. $y_1 = x$, $y_2 = -x$

B. $y_1 = x$, $y_2 = |x|$

C. $y_1 = x$, $y_2 = -|x|$

D. $y_1 = -x$, $y_2 = |x|$

E. $y_1 = |x|$, $y_2 = -|x|$

The answer is C.
The equation for y_1 keeps all the x inputs the same, while the y_2 makes all the outputs negative. Choice C accomplishes these transformations.

CALCULUS

Description of the Examination

The Calculus examination covers skills and concepts that are usually taught in a one-semester college course in calculus. The content of each examination is approximately 60% limits and differential calculus and 40% integral calculus. Algebraic, trigonometric, exponential, logarithmic, and general functions are included. The exam is primarily concerned with an intuitive understanding of calculus and experience with its methods and applications. Knowledge of preparatory mathematics, including algebra, geometry, trigonometry, and analytic geometry is assumed.

The examination contains 44 questions, in two sections, to be answered in approximately 90 minutes. Any time candidates spend on tutorials and providing personal information is in addition to the actual testing time.

- Section 1: 27 questions, approximately 50 minutes.
  No calculator is allowed for this section.

- Section 2: 17 questions, approximately 40 minutes.
  The use of an online graphing calculator (non-CAS) is allowed for this section. Only some of the questions will require the use of the calculator.

Graphing Calculator

A graphing calculator is integrated into the exam software, and it is available to students during Section 2 of the exam.

Only some of the questions actually require the graphing calculator. Students are expected to know how and when to make appropriate use of the calculator. The graphing calculator, together with brief video tutorials, is available to students as a free download for a 30-day trial period. Students are expected to download the calculator and become familiar with its functionality prior to taking the exam.

In order to answer some of the questions in the calculator section of the exam, students may be required to use the online graphing calculator in the following ways:

- Perform calculations (e.g., exponents, roots, trigonometric values, logarithms)
- Graph functions and analyze the graphs
- Find zeros of functions
- Find points of intersection of graphs of functions
- Find minima/maxima of functions
- Find numerical solutions to equations
- Generate a table of values for a function

Knowledge and Skills Required

Questions on the exam require candidates to demonstrate the following abilities:

- Solving routine problems involving the techniques of calculus (approximately 50% of the exam)
- Solving nonroutine problems involving an understanding of the concepts and applications of calculus (approximately 50% of the exam)

CALCULUS

Section I

TIME: 50 Minutes
27 Questions

Directions: Solve each of the following problems without using a calculator. Choose the best answer from those provided. Some questions will require you to enter a numerical answer in the box provided.

Notes:

(1) Figures that accompany questions are intended to provide information useful in answering the questions. All figures lie in a plane unless otherwise indicated. The figures are drawn as accurately as possible EXCEPT when it is stated in a specific question that the figure is not drawn to scale. Straight lines and smooth curves may appear slightly jagged.

(2) Unless otherwise specified, all angles are measured in radians and all numbers used are real numbers.

(3) Unless otherwise specified, the domain of any function f is assumed to be the set of all real numbers x for which $f(x)$ is a real number. The range of f is assumed to be the set of all real numbers $f(x)$, where x is in the domain of f.

(4) In this exam, $\ln(x)$ denotes the natural logarithm of x (the logarithm to the base e).

(5) The inverse of a trigonometric function f may be indicated using the inverse function notation f^{-1} or with the prefix "arc" (e.g., $\sin^{-1}(x) = \arcsin(x)$).

CALCULUS

1. If $y = -3x^2 + 2x + 1$, then $\frac{dy}{dx} =$

   A. $-6x$

   B. $-6x + 1$

   C. $-6x + 2$

   D. $-x^3 + x^2$

   E. $-x^3 + x^2 + x$

2. $\int \sin(2x) dx =$

   A. $-2\cos(2x) + C$

   B. $-\frac{1}{2}\cos(2x) + C$

   C. $\frac{1}{2}\cos(2x) + C$

   D. $2\cos(2x) + C$

   E. $\frac{1}{2}\sin(2x) + C$

3. $\lim\limits_{x \to \infty} \frac{6x^2 + 2x}{5 - 2x^2} =$

   A. -3

   B. -1

   C. 0

   D. $\frac{6}{5}$

   E. The limit does not exist.

4. The velocity of a particle is constantly increasing, as shown in the table below.

$t\ (s)$	0	1	2	3	4
$v\ (m/s)$	0	2	3	5	7

   Using a Riemann sum with four subdivisions, give an upper bound on the distance traveled (in m).

    m

5. Let f be a function defined over all real numbers, and let c be a real number. If $\lim\limits_{x \to c} f(x) = f(c)$, then which of the following statements MUST be true?

   I. f is continuous at $x = c$.
   II. f is differentiable at $x = c$.
   III. f is integrable at $x = c$.

   A. I only

   B. II only

   C. I and II only

   D. I and III only

   E. I, II and III

CALCULUS

6. The graph of a function f is shown in the figure below.

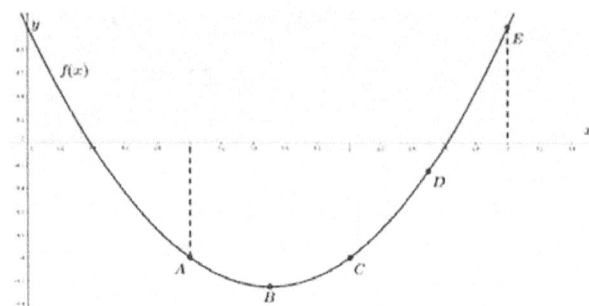

At which of the following points is f' equal to the mean rate of change of f on the interval $[1, 3]$?

A. A

B. B

C. C

D. D

E. E

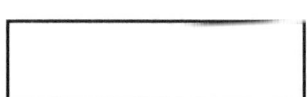

7. If $y = x^2 \cos(x)$, then $\dfrac{d^2y}{dx^2} =$

A. $-2\sin(x)$

B. $-2x\sin(x)$

C. $2\cos(x) - 4x\sin(x) - x^2\cos(x)$

D. $2\cos(x) - 4x\sin(x) + x^2\cos(x)$

E. $2\cos(x) + 4x\sin(x) + x^2\cos(x)$

8. The Riemann sum $\sum_{i=1}^{360} \sin\left(\dfrac{\pi i}{360}\right) \cdot \dfrac{\pi}{360}$ is an approximation for which of the following integrals?

A. $\int_1^{360} \sin(x)\,dx$

B. $\int_1^{360} \sin(\pi x)\,dx$

C. $\int_0^{\pi} \sin(x)\,dx$

D. $\int_0^{\pi} \sin(\pi x)\,dx$

E. $\int_0^{\pi} \sin(360x)\,dx$

9. If $g(x) = \dfrac{x}{2x+1}$, then $g'(x) =$

A. $\dfrac{1}{2}$

B. $\dfrac{4x+1}{2x+1}$

C. $\dfrac{4x+1}{(2x+1)^2}$

D. $\dfrac{1}{2x+1}$

E. $\dfrac{1}{(2x+1)^2}$

CALCULUS

10. Let f and g be differentiable functions on the whole real line, and let $h(x) = f(g(x))$. Use the table of values below to find $h'(1)$.

x	$f(x)$	$f'(x)$	$g(x)$	$g'(x)$
-2	-1	2	-3	0
1	4	5	-2	3

11. If $y = x^x$, then $\frac{dy}{dx} =$

    A. $(\ln(x) + 1)x^x$

    B. $\ln(x)x^x$

    C. x^x

    D. x^{x-1}

    E. $x\ln(x)$

12. Let g be a continuous function on the whole real line, and let $a, b,$ and c be positive constants. $\int_a^b g\left(\frac{x-1}{c}\right) dx$ is equivalent to which of the following integrals?

    A. $\int_{(a-1)/c}^{(b-1)/c} g(u)\, du$

    B. $\frac{1}{c}\int_{(a-1)/c}^{(b-1)/c} g(u)\, du$

    C. $c\int_{(a-1)/c}^{(b-1)/c} g(u)\, du$

    D. $\frac{1}{c}\int_a^b g(u)\, du$

    E. $c\int_a^b g(u)\, du$

13. Let f be differentiable on the whole real line. If $y = -\frac{1}{5}x + \frac{31}{5}$ is normal to the graph of f at $x = 1$, which of the following statements MUST be true?

    I. f is increasing at $x = 1$.
    II. $f'(1) = -\frac{1}{5}$
    III. f is continuous at $x = 1$.

    A. I only

    B. II only

    C. III only

    D. I and II only

    E. I and III only

14. Given $\int_a^b f(x)dx = 4$, $\int_a^c f(x)dx = -2$, and $a < b < c$, evaluate $\int_b^c f(x)dx$.

    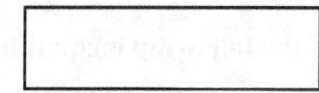

15. If $y = \frac{\sec(x)}{\csc(x)}$, then $\frac{dy}{dx} =$

    A. $-\frac{\sec(x)\tan(x)}{\csc(x)\cot(x)}$

    B. $-\frac{\cos(x)}{\sin(x)}$

    C. $\sec^2(x)$

    D. $-\csc^2(x)$

    E. $\tan(x)$

CALCULUS

16. The limit $\lim_{h \to 0} \frac{(x+h)^2 - 2(x+h) - x^2 + 2x}{h}$ is equal to the derivative of which of the following functions?

    A. $f(x) = x^2$

    B. $f(x) = x^2 - 2x$

    C. $f(x) = x^2 + 2x$

    D. $f(x) = x - 2$

    E. $f(x) = x + 2$

17. The graph of a twice differentiable function f is shown below.

    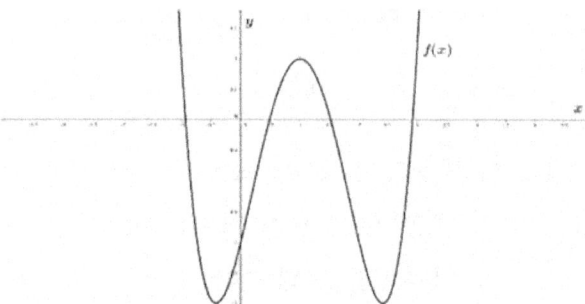

    Which of the following inequalities is true?

    A. $f(1) < f'(1) < f''(1)$

    B. $f(1) < f''(1) < f'(1)$

    C. $f'(1) < f''(1) < f(1)$

    D. $f''(1) < f(1) < f'(1)$

    E. $f''(1) < f'(1) < f(1)$

18. If $y = \arctan(\ln(x))$, then $\frac{dy}{dx} =$

    A. $\frac{1}{(\ln(x))^2 + 1}$

    B. $\frac{1}{(\ln(x))^2 + 1} \cdot \frac{1}{x}$

    C. $\frac{\ln(x)}{(\ln(x))^2 + 1}$

    D. $\frac{1}{\ln(x) + 1}$

    E. $\frac{1}{\ln(x) + 1} \cdot \frac{1}{x}$

19. Below are the graphs of three differentiable functions $f, g,$ and h. Which of the following statements is true?

    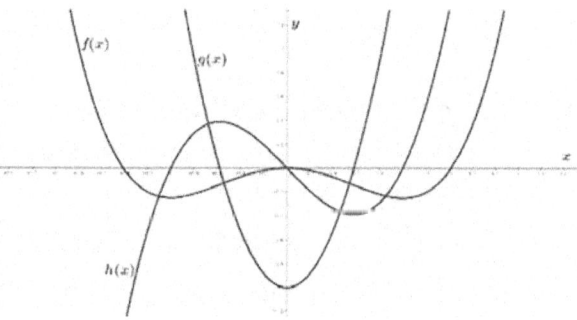

    A. $f'(x) = g(x)$ and $g'(x) = f(x)$

    B. $f'(x) = g(x)$ and $g'(x) = h(x)$

    C. $f'(x) = h(x)$ and $h'(x) = f(x)$

    D. $f'(x) = h(x)$ and $h'(x) = g(x)$

    E. $g'(x) = h(x)$ and $h'(x) = f(x)$

CALCULUS

20. $\int \cot^2(x) dx =$

    A. $-\cot(x) - x + C$

    B. $\cot(x) - x + C$

    C. $\frac{1}{3}\cot^3(x) + C$

    D. $-2\cot^2(x) + C$

    E. $-2\cot^2(x)\csc(x) + C$

21. The graph of $f(t)$ is given below, and $a, b, c, d,$ and e are real numbers.

    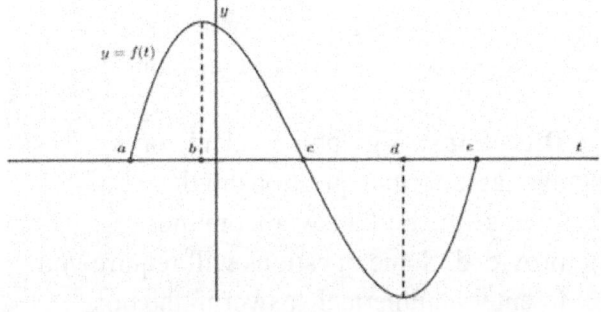

    Let $g(x) = \int_a^x f(t)\, dt$. On which of the following intervals is the graph of g concave up?

    A. (a, b)

    B. (a, c)

    C. $(a, b) \cup (c, d)$

    D. $(a, b) \cup (d, e)$

    E. (b, d)

22. $\int_0^1 \frac{1}{\sqrt{x}} dx =$

    A. 2

    B. 1

    C. $\frac{1}{2}$

    D. 0

    E. Undefined

23. Let $f(x) = 2x^3 + 6x^2 + 6x - 1$. Find all values of x for which $f'(x) = f''(x)$.

    A. 0 and 24

    B. -1 and 1

    C. -2 and 2

    D. $-\sqrt{6}$ and $\sqrt{6}$

    E. -6 and 6

CALCULUS

24. What is the mean value of $f(x) = x^2 + 1$ on the interval $[0, 2]$?

A. 0

B. 1

C. 2

D. $\frac{7}{3}$

E. $\frac{14}{3}$

25. $\lim\limits_{x \to 0} \frac{e^{2x}-1}{\tan(3x)}$

A. $-\infty$

B. 0

C. $\frac{2}{3}$

D. 1

E. ∞

26. Let $f(x)$ be the following piecewise function.

$$f(x) = \begin{cases} x^2 - 1 & x \leq 1 \\ 2x + k & x > 1 \end{cases}$$

Find the value of k for which f is continuous at $x = 1$.

A. -2

B. -1

C. 0

D. 1

E. 2

27. Find the real number b such that $\int_1^b \frac{1}{t} dt = -2$.

A. e^2

B. $e^{1/2}$

C. e^{-2}

D. $\frac{1}{\sqrt{3}}$

E. $-\frac{1}{\sqrt{3}}$

Section II

TIME: 40 Minutes

17 Questions

Directions: A graphing calculator is available for the questions in this section. Choose the best answer from those provided. Some questions will require you to enter a numerical answer in the box provided.

28. Find the slope at the point $(2, 1)$ on the graph of $x^2 - y^2 - x = 1$.

A. $-\frac{3}{2}$

B. -1

C. 0

D. 1

E. $\frac{3}{2}$

156

CALCULUS

29. Find the area of the region bounded by the graphs of $f(x) = (x-1)^3 + 1$ and $g(x) = x$.

   A. 0

   B. $\frac{1}{4}$

   C. $\frac{1}{2}$

   D. $\frac{3}{4}$

   E. 1

30. What is the absolute maximum of the function $f(x) = 2x^3 + 3x^2 - 12x + 4$ on the interval $[0, 2]$?

   A. -3

   B. 4

   C. 8

   D. 24

   E. f has no absolute maximum on $[0,2]$.

31. Find the function $f(x)$ that satisfies both the differential equation $f'(x) = 2x - 3$ and the condition $f(1) = 2$.

   A. $f(x) = x^2 - 3x + 1$

   B. $f(x) = x^2 - 3x + 2$

   C. $f(x) = x^2 - 3x + 3$

   D. $f(x) = x^2 - 3x + 4$

   E. $f(x) = x^2 - 3x + 5$

32. Find the area bounded by the x-axis and the graph of f, where $f(x)$ is the following piecewise function.

$$f(x) = \begin{cases} \frac{1}{2}x + \frac{1}{2} & 0 \leq x < 1 \\ \sqrt{1 - (x-1)^2} & 1 \leq x \leq 2 \end{cases}$$

   A. $1 + \frac{\pi}{2}$

   B. $1 + \frac{\pi}{4}$

   C. $\frac{3}{4} + \frac{\pi}{2}$

   D. $\frac{3}{4} + \frac{\pi}{4}$

   E. $\frac{3}{2}$

33. The displacement s of a particle at time $t \geq 0$ is given by the following function.

$$s(t) = \frac{1}{20}t^5 - \frac{5}{12}t^4 + \frac{4}{3}t^3 - 2t^2$$

Assume that all quantities are in SI units. For what values of t (in seconds) is the acceleration of the particle negative?

   A. $(0,1)$ only

   B. $(0,3)$ only

   C. $(1,2)$ only

   D. $(0,1) \cup (1,2)$

   E. $(0,1) \cup (1,3)$

CALCULUS

34. Use four trapezoids to estimate the area of the region bounded by the graph of $f(x) = \frac{1}{x^4+1}$ and the lines $x = 0, x = 2,$ and $y = 0$. Round your answer to the nearest hundredth.

[ ]

35. $\lim_{x \to 0^+} x^{\sin(x)} =$

A. -1

B. 0

C. $\frac{1}{2}$

D. 1

E. 2

36. The acceleration a (in $\frac{m}{s^2}$) of a particle at time $t \geq 0$ (in s) is given by $a(t) = te^{-t^2}$. At $t = 0$ s, the velocity of the particle is $2.5 \frac{m}{s}$. What is the velocity (in $\frac{m}{s}$) of the particle at $t = 2$ s?

A. $2 - \frac{1}{2e^4}$

B. $2 + \frac{1}{2e^4}$

C. $3 - \frac{1}{2e^4}$

D. 3

E. $3 + \frac{1}{2e^4}$

37. Find the equation of the line that passes through the point $(3, 4)$ and that, together with the $x-$ and $y-$axes, forms a triangular region in the first quadrant of minimum area.

A. $3x + 4y - 25 = 0$

B. $3x - 4y + 7 = 0$

C. $4x + 3y - 24 = 0$

D. $4x - 3y = 0$

E. $x + y - 7 = 0$

38. Which of the following inequalities is true?

Let $A = \int_{-1/2}^{1/2} f(x)\, dx$, $B = \int_{-1/2}^{1/2} g(x)\, dx$, and $C = \int_{-1/2}^{1/2} h(x)\, dx$, where $f(x) = \frac{1}{2(x^2+1)}, g(x) = \frac{x^2}{x^2+1},$ and $h(x) = e^{-x^2}$

A. $A < B < C$

B. $A < C < B$

C. $B < A < C$

D. $B < C < A$

E. $C < A < B$

CALCULUS

39. Let $f(x) = -\frac{1}{3}x^3 - x + 1$. Find any values of x in the interval $[-3, 3]$ at which the instantaneous rate of change of f equals the average rate of change of f on $[-3, 3]$.

   A. -3

   B. $\pm\sqrt{5}$

   C. $\pm\sqrt{3}$

   D. $\pm\sqrt{2}$

   E. 3

40. Consider the region bounded by the graph of $y = 4 - 2x^2$ and the x-axis. Find the area of the largest isosceles triangle that can be inscribed in this region with one vertex at the origin and the base parallel to the x-axis. Round your answer to the nearest hundredth.

   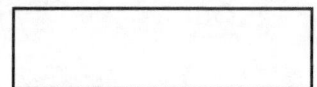

41. Which of the following functions is strictly monotonic on its entire domain?

   A. $f(x) = x^3 - 2x$

   B. $f(x) = x^3 - x$

   C. $f(x) = x^3 + x$

   D. $f(x) = x^2 - x$

   E. $f(x) = x^2 + x$

42. The radius of a circle is increasing at a constant rate of $3\ \frac{cm}{s}$. At what rate (in $\frac{cm^2}{s}$) is the area of the circle increasing when the radius is $5\ cm$?

   A. 6π

   B. 9π

   C. 10π

   D. 25π

   E. 30π

43. $\lim\limits_{x \to \pi/2} \sec(x)\tan(x) =$

   A. $-\infty$

   B. 0

   C. 1

   D. ∞

   E. The limit does not exist.

44. Find all real numbers c in $[a, b]$ such that $\int_a^b f(x)\, dx = f(c)(b-a)$ if $a = 0, b = 2,$ and $f(x) = 3x^2 + 1$.

A. $\pm \frac{\sqrt{3}}{3}$

B. $\pm \frac{2\sqrt{3}}{3}$

C. $\pm \frac{4}{3}$

D. $\frac{2\sqrt{3}}{3}$

E. $\frac{4}{3}$

CALCULUS

ANSWER KEY

Question Number	Correct Answer	Your Answer
1	C	
2	B	
3	A	
4	17	
5	D	
6	C	
7	C	
8	C	
9	E	
10	6	
11	A	
12	C	
13	E	
14	6	
15	C	

Question Number	Correct Answer	Your Answer
16	B	
17	E	
18	B	
19	D	
20	A	
21	D	
22	A	
23	B	
24	D	
25	C	
26	A	
27	C	
28	E	
29	C	
30	C	

Question Number	Correct Answer	Your Answer
31	D	
32	D	
33	A	
34	1.07	
35	D	
36	C	
37	C	
38	C	
39	C	
40	2.18	
41	C	
42	E	
43	D	
44	D	

CALCULUS

RATIONALES

Section I

TIME: 50 Minutes
27 Questions

Directions: Solve each of the following problems without using a calculator. Choose the best answer from those provided. Some questions will require you to enter a numerical answer in the box provided.

Notes:

(1) Figures that accompany questions are intended to provide information useful in answering the questions. All figures lie in a plane unless otherwise indicated. The figures are drawn as accurately as possible EXCEPT when it is stated in a specific question that the figure is not drawn to scale. Straight lines and smooth curves may appear slightly jagged.

(2) Unless otherwise specified, all angles are measured in radians and all numbers used are real numbers.

(3) Unless otherwise specified, the domain of any function f is assumed to be the set of all real numbers x for which $f(x)$ is a real number. The range of f is assumed to be the set of all real numbers $f(x)$, where x is in the domain of f.

(4) In this exam, $\ln(x)$ denotes the natural logarithm of x (the logarithm to the base e).

(5) The inverse of a trigonometric function f may be indicated using the inverse function notation f^{-1} or with the prefix "arc" (e.g., $\sin^{-1}(x) = \arcsin(x)$).

CALCULUS

1. If $y = -3x^2 + 2x + 1$, then $\frac{dy}{dx} =$

    A. $-6x$

    B. $-6x + 1$

    C. $-6x + 2$

    D. $-x^3 + x^2$

    E. $-x^3 + x^2 + x$

The answer is C.
Differentiate term by term and use the Power Rule for derivatives.

2. $\int \sin(2x) dx =$

    A. $-2\cos(2x) + C$

    B. $-\frac{1}{2}\cos(2x) + C$

    C. $\frac{1}{2}\cos(2x) + C$

    D. $2\cos(2x) + C$

    E. $\frac{1}{2}\sin(2x) + C$

The answer is B.
Let $u = 2x$. Then $du = 2\, dx \implies dx = \frac{1}{2} du$

$$\int \sin(2x) dx = \frac{1}{2} \int \sin(u)\, du$$

$$\int \sin(2x) dx = -\frac{1}{2}\cos(u) + C$$

$$\int \sin(2x) dx = -\frac{1}{2}\cos(2x) + C$$

CALCULUS

3. $\lim\limits_{x \to \infty} \dfrac{6x^2+2x}{5-2x^2} =$

    A. -3

    B. -1

    C. 0

    D. $\dfrac{6}{5}$

    E. The limit does not exist.

The answer is A.

Divide the numerator and denominator by x^2.

$$\lim_{x \to \infty} \frac{6x^2+2x}{5-2x^2} = \lim_{x \to \infty} \frac{\frac{6x^2}{x^2}+\frac{2x}{x^2}}{\frac{5}{x^2}-\frac{2x^2}{x^2}}$$

$$\lim_{x \to \infty} \frac{6x^2+2x}{5-2x^2} = \lim_{x \to \infty} \frac{6+\frac{2}{x}}{\frac{5}{x^2}-2}$$

$$\lim_{x \to \infty} \frac{6x^2+2x}{5-2x^2} = \frac{6+0}{0-2}$$

$$\lim_{x \to \infty} \frac{6x^2+2x}{5-2x^2} = -3$$

CALCULUS

4. The velocity of a particle is constantly increasing, as shown in the table below.

t (s)	0	1	2	3	4
v (m/s)	0	2	3	5	7

Using a Riemann sum with four subdivisions, give an upper bound on the distance traveled (in m).

The answer is 17.
A sketch of the data points and the rectangles for the upper sum is shown below.

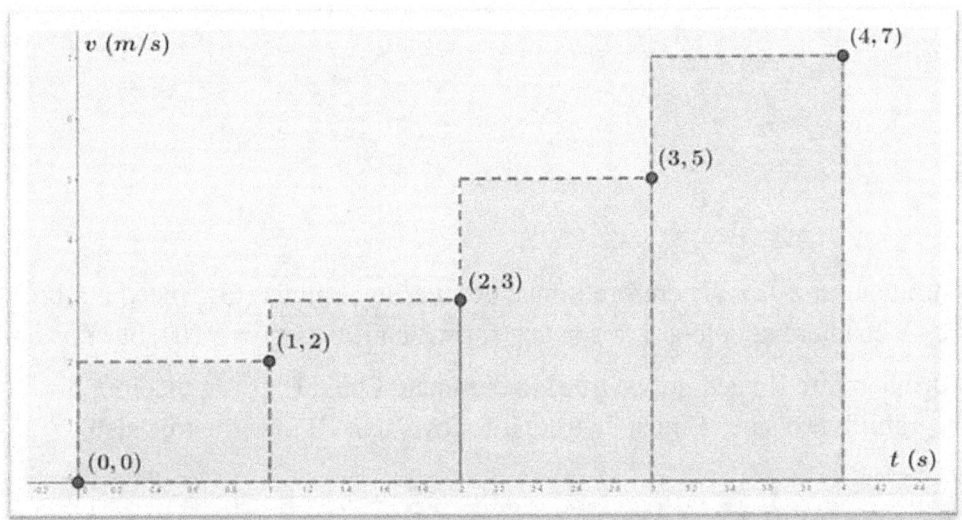

The upper bound is the sum of the areas of the four rectangles.

$$Upper\ Bound = \left(2\ \frac{m}{s}\right)(1\ s) + \left(3\ \frac{m}{s}\right)(1\ s) + \left(5\ \frac{m}{s}\right)(1\ s) + \left(7\ \frac{m}{s}\right)(1\ s)$$
$$Upper\ Bound = 2m + 3m + 5m + 7m$$
$$Upper\ Bound = 17\ m$$

CALCULUS

5. Let f be a function defined over all real numbers, and let c be a real number. If $\lim_{x \to c} f(x) = f(c)$, then which of the following statements MUST be true?

   I. f is continuous at $x = c$.
   II. f is differentiable at $x = c$.
   III. f is integrable at $x = c$.

   A. I only

   B. II only

   C. I and II only

   D. I and III only

   E. I, II and III

The answer is D.
By definition, f is continuous at $x = c$. Therefore I must be true, so eliminate B. f need not be differentiable at $x = c$. A counterexample is $f(x) = |x|$, for which $\lim_{x \to 0} f(x) = f(0)$, but f is not differentiable at $x = 0$. Therefore II need not be true, so eliminate C and E. It is a theorem that continuity implies integrability, so since I must be true, it follows that III must be true also. Eliminate A and choose D.

CALCULUS

6. The graph of a function f is shown in the figure below.

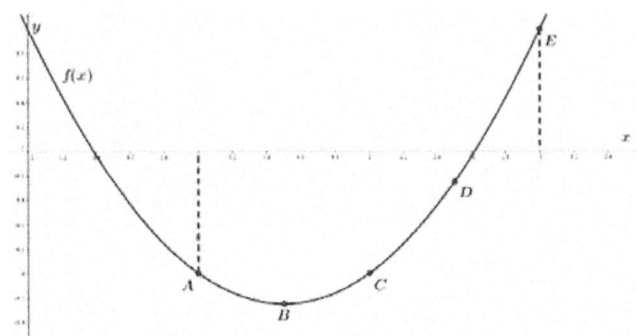

At which of the following points is f' equal to the mean rate of change of f on the interval $[1, 3]$?

A. A

B. B

C. C

D. D

E. E

The answer is C.

We are looking for the point at which the line tangent to the graph of f is *parallel to* the line secant to the graph on $[1,3]$. The graph with the tangent and secant lines is shown below.

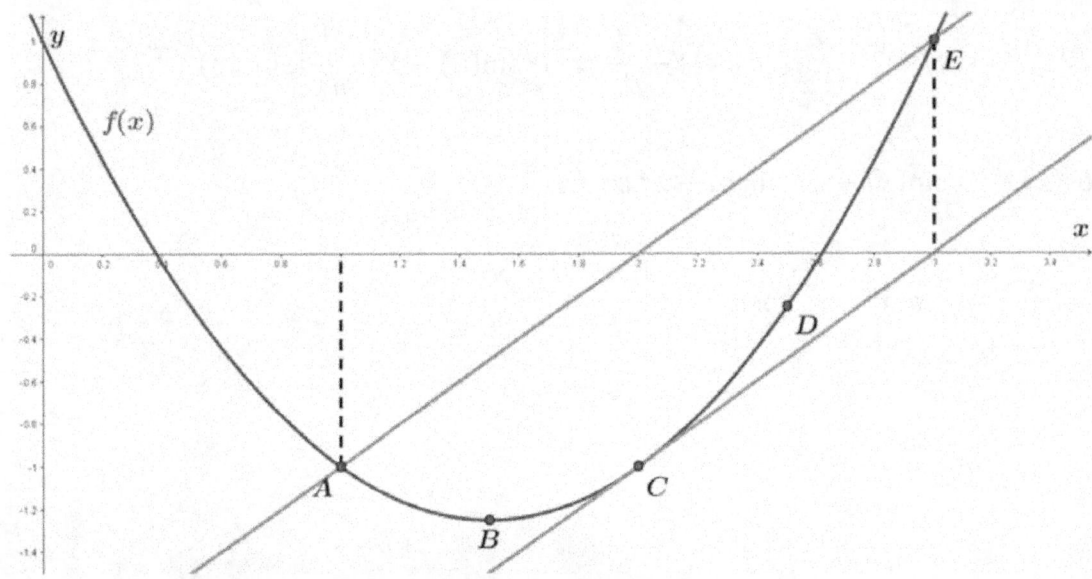

CALCULUS

7. If $y = x^2 \cos(x)$, then $\frac{d^2y}{dx^2} =$

    A. $-2 \sin(x)$

    B. $-2x \sin(x)$

    C. $2 \cos(x) - 4x \sin(x) - x^2 \cos(x)$

    D. $2 \cos(x) - 4x \sin(x) + x^2 \cos(x)$

    E. $2 \cos(x) + 4x \sin(x) + x^2 \cos(x)$

The answer is C.

First use the Product Rule to obtain $\frac{dy}{dx}$.

$$\frac{dy}{dx} = \frac{d}{dx}(x^2) \cdot \cos(x) + x^2 \cdot \frac{d}{dx}(\cos(x))$$

$$\frac{dy}{dx} = 2x \cos(x) - x^2 \sin(x)$$

Now differentiate term by term to obtain $\frac{d^2y}{dx^2}$, again heeding the Product Rule.

$$\frac{d^2y}{dx^2} = \frac{d}{dx}(2x \cos(x)) - \frac{d}{dx}(x^2 \sin(x))$$

$$\frac{d^2y}{dx^2} = \frac{d}{dx}(2x) \cdot \cos(x) + 2x \cdot \frac{d}{dx}(\cos(x)) - \frac{d}{dx}(x^2) \cdot \sin(x) - x^2 \cdot \frac{d}{dx}(\sin(x))$$

$$\frac{d^2y}{dx^2} = 2 \cos(x) - 2x \sin(x) - 2x \sin(x) - x^2 \cos(x)$$

$$\frac{d^2y}{dx^2} = 2 \cos(x) - 4x \sin(x) - x^2 \cos(x)$$

CALCULUS

8. The Riemann sum $\sum_{i=1}^{360} \sin\left(\frac{\pi i}{360}\right) \cdot \frac{\pi}{360}$ is an approximation for which of the following integrals?

A. $\int_1^{360} \sin(x)\, dx$

B. $\int_1^{360} \sin(\pi x)\, dx$

C. $\int_0^{\pi} \sin(x)\, dx$

D. $\int_0^{\pi} \sin(\pi x)\, dx$

E. $\int_0^{\pi} \sin(360x)\, dx$

The answer is C.
The general form of a Riemann sum for a function $f(x)$ is as follows.

$$\sum_{i=1}^{n} f(c_i)\, \Delta x_i$$

Comparing this to the given Riemann sum, we see that $n = 360$, $\Delta x_i = \frac{\pi}{360}$, $c_i = \frac{\pi i}{360} = i\Delta x_i$, and $f(x) = \sin(x)$. Eliminate B, D, and E, as they have the wrong integrand. Since Δx_i is a constant, it defines a *regular partition* of the interval of integration. In general, for a regular partition of an interval $[a, b]$ into n subintervals of width Δx, we have $\Delta x = \frac{b-a}{n}$. Comparing this to our expression for Δx_i, we see that $b - a = \pi$. This is the width of the interval of integration. Eliminate A, and choose C.

9. If $g(x) = \frac{x}{2x+1}$, then $g'(x) =$

A. $\frac{1}{2}$

B. $\frac{4x+1}{2x+1}$

C. $\frac{4x+1}{(2x+1)^2}$

D. $\frac{1}{2x+1}$

E. $\frac{1}{(2x+1)^2}$

The answer is E.
Use the Quotient Rule.

$$g'(x) = \frac{\frac{d}{dx}(x) \cdot (2x+1) - x \cdot \frac{d}{dx}(2x+1)}{(2x+1)^2}$$

$$g'(x) = \frac{(1)(2x+1) - x(2)}{(2x+1)^2}$$

$$g'(x) = \frac{2x+1-2x}{(2x+1)^2}$$

$$g'(x) = \frac{1}{(2x+1)^2}$$

CALCULUS

10. Let f and g be differentiable functions on the whole real line, and let $h(x) = f(g(x))$. Use the table of values below to find $h'(1)$.

x	$f(x)$	$f'(x)$	$g(x)$	$g'(x)$
-2	-1	2	-3	0
1	4	5	-2	3

The answer is 6.
Use the Chain Rule.
$h'(x) = f'(g(x))g'(x)$
$h'(1) = f'(g(1))g'(1)$
$h'(1) = f'(-2)g'(1)$
$h'(1) = (2)(3)$
$h'(1) = 6$

11. If $y = x^x$, then $\frac{dy}{dx} =$

   A. $(\ln(x) + 1)x^x$

   B. $\ln(x)x^x$

   C. x^x

   D. x^{x-1}

   E. $x \ln(x)$

The answer is A.
Use Logarithmic Differentiation.

$y = x^x$
$\ln(y) = \ln(x^x)$
$\ln(y) = x \ln(x)$

$\frac{d}{dx}(\ln(y)) = \frac{d}{dx}(x) \cdot \ln(x) + x \cdot \frac{d}{dx}(\ln(x))$

$\frac{y'}{y} = 1 \cdot \ln(x) + x \cdot \frac{1}{x}$

$y' = (\ln(x) + 1)y$
$y' = (\ln(x) + 1)x^x$

CALCULUS

12. Let g be a continuous function on the whole real line, and let a, b, and c be positive constants. $\int_a^b g\left(\frac{x-1}{c}\right) dx$ is equivalent to which of the following integrals?

    A. $\int_{(a-1)/c}^{(b-1)/c} g(u)\, du$

    B. $\frac{1}{c}\int_{(a-1)/c}^{(b-1)/c} g(u)\, du$

    C. $c\int_{(a-1)/c}^{(b-1)/c} g(u)\, du$

    D. $\frac{1}{c}\int_a^b g(u)\, du$

    E. $c\int_a^b g(u)\, du$

The answer is C.

Let $u = \frac{x-1}{c}$. Then $du = \frac{1}{c}dx \Rightarrow dx = c\, du$. Converting the limits of integration, we obtain the following new limits of integration.

$$x = a \Rightarrow u = \frac{a-1}{c}$$

$$x = b \Rightarrow u = \frac{b-1}{c}$$

So the integral becomes $c\int_{(a-1)/c}^{(b-1)/c} g(u)\, du$.

CALCULUS

13. Let f be differentiable on the whole real line. If $y = -\frac{1}{5}x + \frac{31}{5}$ is normal to the graph of f at $x = 1$, which of the following statements MUST be true?

    I. f is increasing at $x = 1$.
    II. $f'(1) = -\frac{1}{5}$
    III. f is continuous at $x = 1$.

    A. I only

    B. II only

    C. III only

    D. I and II only

    E. I and III only

The answer is E.
The slope m_n of the line normal to the graph of f at $x = 1$ is $-\frac{1}{5}$. Since the line tangent to the graph of f at $x = 1$ is *perpendicular* to the normal line, its slope is $m_t = -\frac{1}{m_n} = 5$. This implies that $f'(1) = 5$, and since this value is positive, f is increasing at $x = 1$. Therefore, I is true and II is false. Eliminate B, C, and D. Since differentiability implies continuity, f is continuous at $x = 1$. Therefore III is true. Eliminate A, and choose E.

CALCULUS

14. Given $\int_a^b f(x)dx = 4$, $\int_a^c f(x)dx = -2$, and $a < b < c$, evaluate $\int_b^c f(x)dx$.

   A. $-\dfrac{\sec(x)\tan(x)}{\csc(x)\cot(x)}$

   B. $-\dfrac{\cos(x)}{\sin(x)}$

   C. $\sec^2(x)$

   D. $-\csc^2(x)$

   E. $\tan(x)$

The answer is 6.

$$\int_a^b f(x)\,dx = \int_a^c f(x)\,dx + \int_c^b f(x)\,dx$$

$$4 = -2 + \int_c^b f(x)\,dx$$

$$6 = \int_c^b f(x)\,dx$$

CALCULUS

15. If $y = \dfrac{\sec(x)}{\csc(x)}$, then $\dfrac{dy}{dx} =$

   A. $-\dfrac{\sec(x)\tan(x)}{\csc(x)\cot(x)}$

   B. $-\dfrac{\cos(x)}{\sin(x)}$

   C. $\sec^2(x)$

   D. $-\csc^2(x)$

   E. $\tan(x)$

The answer is C.
First simplify the function.

$$y = \dfrac{\dfrac{1}{\cos(x)}}{\dfrac{1}{\sin(x)}}$$

$$y = \dfrac{\sin(x)}{\cos(x)}$$

$$y = \tan(x)$$

Now take the derivative. $\dfrac{dy}{dx} = \sec^2(x)$

CALCULUS

16. The limit $\lim_{h \to 0} \frac{(x+h)^2 - 2(x+h) - x^2 + 2x}{h}$ is equal to the derivative of which of the following functions?

   A. $f(x) = x^2$

   B. $f(x) = x^2 - 2x$

   C. $f(x) = x^2 + 2x$

   D. $f(x) = x - 2$

   E. $f(x) = x + 2$

The answer is B.
If $f(x) = x^2 - 2x$, then by definition $f'(x) = \lim_{h \to 0} \frac{(x+h)^2 - 2(x+h) - x^2 + 2x}{h}$.

17. The graph of a twice differentiable function f is shown below.

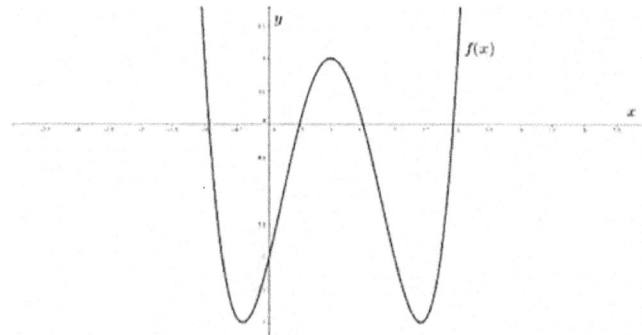

   Which of the following inequalities is true?

   A. $f(1) < f'(1) < f''(1)$

   B. $f(1) < f''(1) < f'(1)$

   C. $f'(1) < f''(1) < f(1)$

   D. $f''(1) < f(1) < f'(1)$

   E. $f''(1) < f'(1) < f(1)$

The answer is E.
From the graph, we can see that $f(1) = 1$. We can also see that f has a relative maximum at $x = 1$, and since f is differentiable, this implies $f'(1) = 0$. So $f'(1) < f(1)$, which means we can

CALCULUS

eliminate A, B, and D. We cannot tell the value of $f''(1)$, but since the graph of f is *concave down* at $x = 1$, it follows that $f''(1) < 0$. Eliminate C and choose E.

18. If $y = \arctan(\ln(x))$, then $\frac{dy}{dx} =$

A. $\dfrac{1}{(\ln(x))^2+1}$

B. $\dfrac{1}{(\ln(x))^2+1} \cdot \dfrac{1}{x}$

C. $\dfrac{\ln(x)}{(\ln(x))^2+1}$

D. $\dfrac{1}{\ln(x)+1}$

E. $\dfrac{1}{\ln(x)+1} \cdot \dfrac{1}{x}$

The answer is B.
Use the Chain Rule.

$$\frac{dy}{dx} = \frac{1}{(\ln(x))^2 + 1} \cdot \frac{d}{dx}(\ln(x))$$

$$\frac{dy}{dx} = \frac{1}{(\ln(x))^2 + 1} \cdot \frac{1}{x}$$

CALCULUS

19. Below are the graphs of three differentiable functions $f, g,$ and h. Which of the following statements is true?

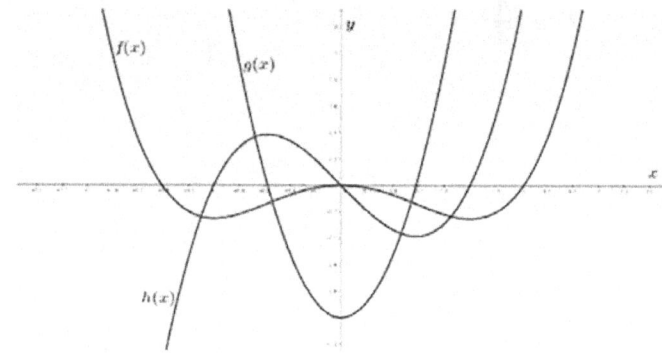

A. $f'(x) = g(x)$ and $g'(x) = f(x)$

B. $f'(x) = g(x)$ and $g'(x) = h(x)$

C. $f'(x) = h(x)$ and $h'(x) = f(x)$

D. $f'(x) = h(x)$ and $h'(x) = g(x)$

E. $g'(x) = h(x)$ and $h'(x) = f(x)$

The answer is D.

f has two zeros (one positive and one negative) that do not correspond to relative extrema of either g or h, so f cannot be the derivative of either of these two functions. Eliminate A, C, and E. None of the relative extrema of f correspond to zeros of g, so g cannot be the derivative of f. Eliminate B and choose D.

CALCULUS

20. $\int \cot^2(x)\, dx =$

   A. $-\cot(x) - x + C$

   B. $\cot(x) - x + C$

   C. $\frac{1}{3}\cot^3(x) + C$

   D. $-2\cot^2(x) + C$

   E. $-2\cot^2(x)\csc(x) + C$

The answer is A.
Use the identity $\cot^2(x) = \csc^2(x) - 1$.

$$\int \cot^2(x)\, dx = \int \left(\csc^2(x) - 1\right) dx$$

$$\int \cot^2(x)\, dx = -\cot(x) - x + C$$

CALCULUS

21. The graph of $f(t)$ is given below, and $a, b, c, d,$ and e are real numbers.

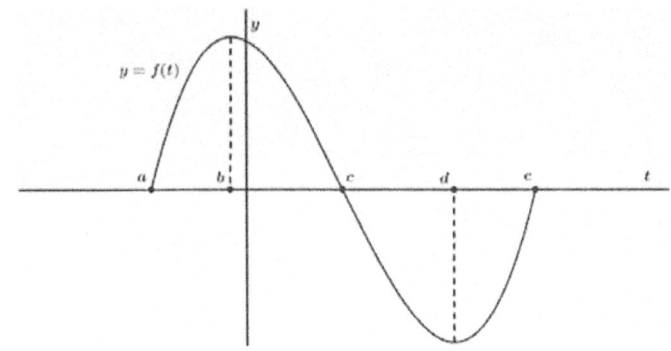

Let $g(x) = \int_a^x f(t)\, dt$. On which of the following intervals is the graph of g concave up?

A. (a, b)

B. (a, c)

C. $(a, b) \cup (c, d)$

D. $(a, b) \cup (d, e)$

E. (b, d)

The answer is D.
Use the Second Fundamental Theorem of Calculus.

$g'(x) = f(x)$
$g''(x) = f'(x)$

The graph of g is concave up on every interval for which $g''(x) > 0$, or $f'(x) > 0$. So the graph of g is concave up on every interval on which f is *increasing*. From the graph, we see that f is increasing for all t (or x) in either (a, b) or (d, e).

CALCULUS

22. $\int_0^1 \frac{1}{\sqrt{x}} dx =$

    A. 2

    B. 1

    C. $\frac{1}{2}$

    D. 0

    E. Undefined

The answer is A.

Because $\frac{1}{\sqrt{x}} \to \infty$ as $x \to 0^+$, this is an improper integral, so we'll handle it accordingly.

$$\int_0^1 \frac{1}{\sqrt{x}} dx = \lim_{a \to 0^+} \int_a^1 \frac{1}{\sqrt{x}} dx$$

$$\int_0^1 \frac{1}{\sqrt{x}} dx = \lim_{a \to 0^+} \int_a^1 x^{-1/2} dx$$

$$\int_0^1 \frac{1}{\sqrt{x}} dx = 2 \lim_{a \to 0^+} x^{1/2} \Big|_a^1$$

$$\int_0^1 \frac{1}{\sqrt{x}} dx = 2 \lim_{a \to 0^+} \left(1^{1/2} - a^{1/2}\right)$$

$$\int_0^1 \frac{1}{\sqrt{x}} dx = 2$$

CALCULUS

23. Let $f(x) = 2x^3 + 6x^2 + 6x - 1$. Find all values of x for which $f'(x) = f''(x)$.

   A. 0 and 24

   B. -1 and 1

   C. -2 and 2

   D. $-\sqrt{6}$ and $\sqrt{6}$

   E. -6 and 6

The answer is B.
$f'(x) = 6x^2 + 12x + 6$
$f''(x) = 12x + 12$

$f'(x) = f''(x)$
$6x^2 + 12x + 6 = 12x + 12$
$6x^2 - 6 = 0$
$6(x-1)(x+1) = 0$
$x = \pm 1$

CALCULUS

24. What is the mean value of $f(x) = x^2 + 1$ on the interval $[0, 2]$?

    A. 0

    B. 1

    C. 2

    D. $\frac{7}{3}$

    E. $\frac{14}{3}$

The answer is D.
Let $\bar{f}$ the average value of f on $[0,2]$.

$$\bar{f} = \frac{1}{2-0} \int_0^2 (x^2 + 1)\, dx$$

$$\bar{f} = \frac{1}{2}\left(\frac{1}{3}x^3 + x\right)\Big|_0^2$$

$$\bar{f} = \frac{1}{2}\left[\left(\frac{1}{3}(2)^3 + 2\right) - \left(\frac{1}{3}(0)^3 + 0\right)\right]$$

$$\bar{f} = \frac{7}{3}$$

CALCULUS

25. $\lim\limits_{x \to 0} \dfrac{e^{2x}-1}{\tan(3x)}$

   A. $-\infty$

   B. 0

   C. $\dfrac{2}{3}$

   D. 1

   E. ∞

The answer is C.
Direct substitution yields the following.

$$\lim_{x \to 0} \frac{e^{2x}-1}{\tan(3x)} = \frac{e^0 - 1}{\tan(0)} = \frac{0}{0}$$

Since we have obtained the indeterminate form $\dfrac{0}{0}$, we may apply L'Hôpital's rule.

$$\lim_{x \to 0} \frac{e^{2x}-1}{\tan(3x)} = \lim_{x \to 0} \frac{\frac{d}{dx}(e^{2x}-1)}{\frac{d}{dx}(\tan(3x))}$$

$$\lim_{x \to 0} \frac{e^{2x}-1}{\tan(3x)} = \lim_{x \to 0} \frac{2e^{2x}}{3\sec^2(3x)}$$

$$\lim_{x \to 0} \frac{e^{2x}-1}{\tan(3x)} = \frac{2e^0}{3\sec^2(0)}$$

$$\lim_{x \to 0} \frac{e^{2x}-1}{\tan(3x)} = \frac{2}{3}$$

CALCULUS

26. Let $f(x)$ be the following piecewise function.

$$f(x) = \begin{cases} x^2 - 1 & x \leq 1 \\ 2x + k & x > 1 \end{cases}$$

Find the value of k for which f is continuous at $x = 1$.

A. -2

B. -1

C. 0

D. 1

E. 2

The answer is A.
The condition that f is continuous at $x = 1$ may be stated as $\lim\limits_{x \to 1} f(x) = f(1)$. In order for this limit to exist, both the left- and right-handed limits must exist, and they must be equal to each other. This requirement gives us the following.

$$\lim_{x \to 1^-} f(x) = \lim_{x \to 1^+} f(x)$$

$$\lim_{x \to 1^-} (x^2 - 1) = \lim_{x \to 1^+} (2x + k)$$

$1^2 - 1 = 2(1) + k$
$0 = 2 + k$
$k = -2$

CALCULUS

27. Find the real number b such that $\int_1^b \frac{1}{t} dt = -2$.

A. e^2

B. $e^{1/2}$

C. e^{-2}

D. $\frac{1}{\sqrt{3}}$

E. $-\frac{1}{\sqrt{3}}$

The answer is C.
The function $\ln(x)$ is *defined* as $\ln(x) = \int_1^x \frac{1}{t} dt$. So we can rewrite the equation in the problem statement as follows.

$\ln(b) = -2$
$b = e^{-2}$

CALCULUS

28. Find the slope at the point $(2, 1)$ on the graph of $x^2 - y^2 - x = 1$.

A. $-\frac{3}{2}$

B. -1

C. 0

D. 1

E. $\frac{3}{2}$

The answer is E.
Use implicit differentiation to find y' in terms of x and y.

$$\frac{d}{dx}(x^2 - y^2 - x) = \frac{d}{dx}(1)$$

$$2x - 2yy' - 1 = 0$$

$$-2yy' = -2x + 1$$

$y' = \frac{-2x+1}{-2y}$ The slope m of the graph at the point $(2,1)$ is found by evaluating y' at that point.

$$m = \left.\frac{-2x + 1}{-2y}\right|_{(2,1)}$$

$$m = \frac{-2(2) + 1}{-2(1)}$$

$m = \frac{3}{2}$ The graph of the equation and its tangent line at $(2,1)$ is shown below.

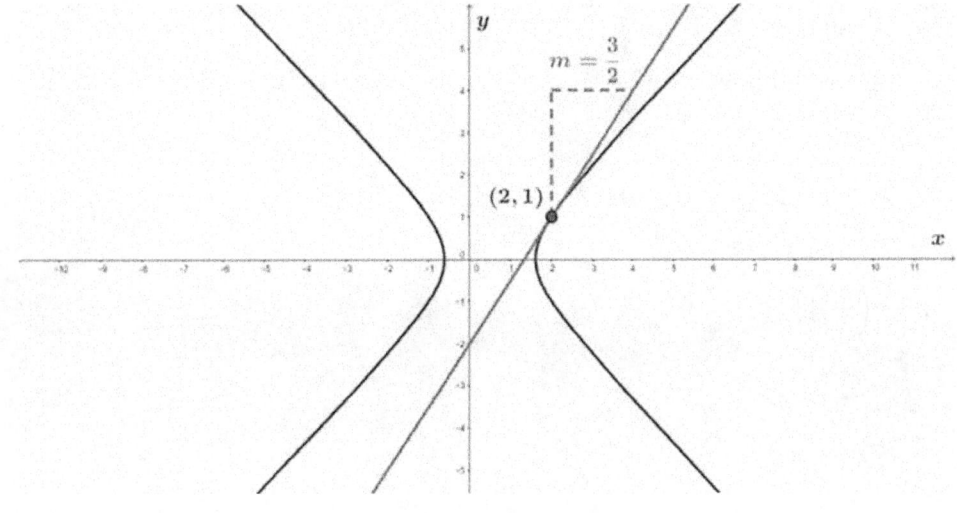

CALCULUS

29. Find the area of the region bounded by the graphs of $f(x) = (x-1)^3 + 1$ and $g(x) = x$.

   A. 0

   B. $\frac{1}{4}$

   C. $\frac{1}{2}$

   D. $\frac{3}{4}$

   E. 1

The answer is C.
It would be helpful to look at a sketch of the graphs before we get started.

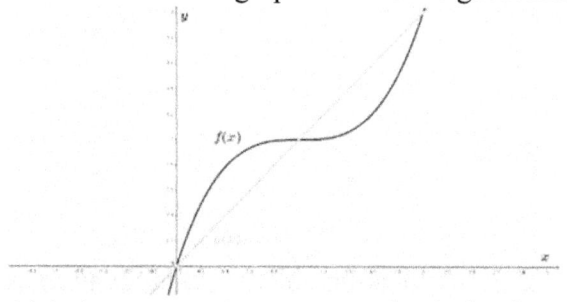

There are three points at which the two graphs intersect: $(0,0)$, $(1,1)$, and $(2,2)$. On the interval $(0,1)$, $f(x) > g(x)$, and on the interval $(1,2)$, $f(x) < g(x)$. We must account for this when we integrate to find the area A of the region.

$$A = \int_0^1 [f(x) - g(x)]\, dx + \int_1^2 [g(x) - f(x)]\, dx$$

$$A = \int_0^1 [(x-1)^3 + 1 - x]\, dx + \int_1^2 [x - (x-1)^3 - 1]\, dx$$

$$A = \int_0^1 (x^3 - 3x^2 + 2x)\, dx + \int_1^2 (-x^3 + 3x^2 - 2x)\, dx$$

$$A = \left(\frac{1}{4}x^4 - x^3 + x^2\right)\Big|_0^1 + \left(-\frac{1}{4}x^4 + x^3 - x^2\right)\Big|_1^2$$

$$A = \left(\frac{1}{4} - 0\right) + \left(0 - \left(-\frac{1}{4}\right)\right)$$

$$A = \frac{1}{2}$$

CALCULUS

30. What is the absolute maximum of the function $f(x) = 2x^3 + 3x^2 - 12x + 4$ on the interval $[0, 2]$?

    A. -3

    B. 4

    C. 8

    D. 24

    E. f has no absolute maximum on $[0,2]$.

The answer is C.
Since f is continuous on $[0,2]$, it has an absolute maximum on $[0,2]$. Eliminate E. The absolute maximum will occur either at one of the endpoints of the interval or at a critical number of f inside the interval. Find the critical numbers of f.

$f'(x) = 6x^2 + 6x - 12 = 0$
$6(x^2 + x - 2) = 0$
$6(x + 2)(x - 1) = 0$
$x = -2, 1$

Since $x = -2$ is not in the interval $[0,2]$, we need not consider it. We find the absolute maximum by evaluating f at the critical number $x = 1$ and at the endpoints $x = 0, 2$.

$f(0) = 2(0)^3 + 3(0)^2 - 12(0) + 4 = 4$
$f(1) = 2(1)^3 + 3(1)^2 - 12(1) + 4 = -3$
$f(2) = 2(2)^3 + 3(2)^2 - 12(2) + 4 = 8$

The graph of f is shown below with the relevant points labeled.

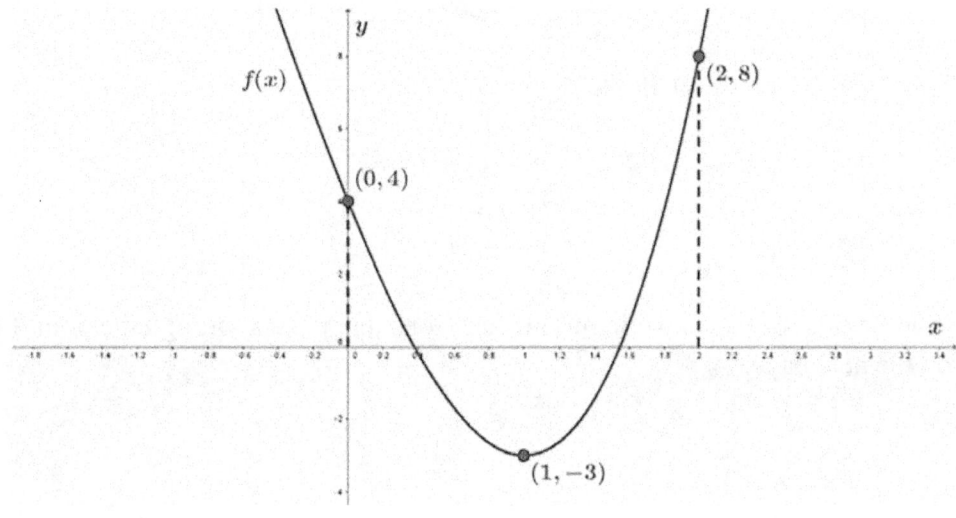

CALCULUS

31. Find the function $f(x)$ that satisfies both the differential equation $f'(x) = 2x - 3$ and the condition $f(1) = 2$.

    A. $f(x) = x^2 - 3x + 1$

    B. $f(x) = x^2 - 3x + 2$

    C. $f(x) = x^2 - 3x + 3$

    D. $f(x) = x^2 - 3x + 4$

    E. $f(x) = x^2 - 3x + 5$

The answer is D.
First solve the differential equation by integration.
$$f(x) = \int (2x - 3)\, dx$$

$$f(x) = x^2 - 3x + C$$

So far we have a family of functions parameterized by the real constant C. Some representative members of this family are graphed below.

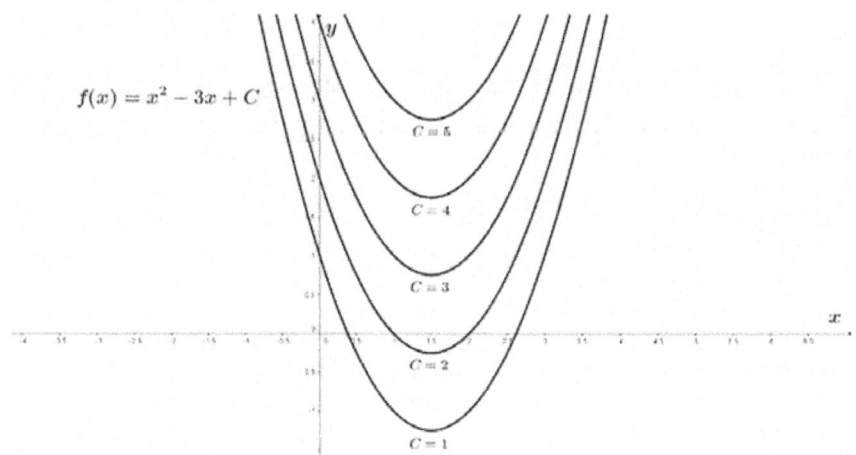

We will find the constant C by applying the condition $f(1) = 2$.
$$f(1) = 2$$
$$(1)^2 - 3(1) + C = 2$$
$$-2 + C = 2$$
$$C = 4$$
So $f(x) = x^2 - 3x + 4$.

This is the one member of the above family of functions that passes through the point (1,2), as illustrated on the next page.

CALCULUS

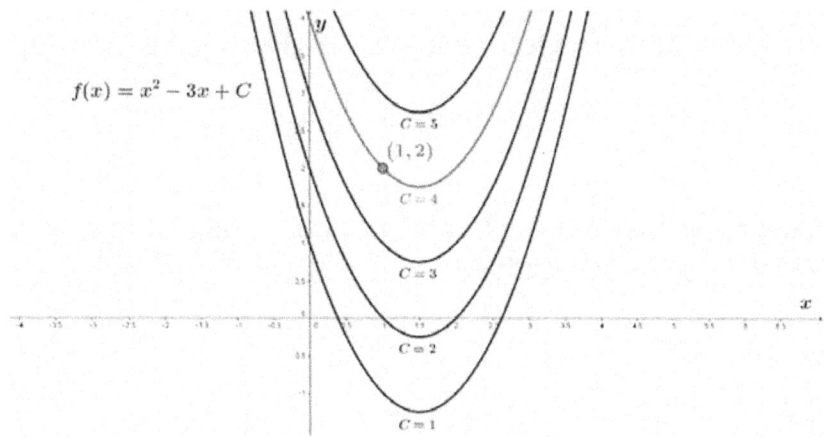

32. Find the area bounded by the x-axis and the graph of f, where $f(x)$ is the following piecewise function.

$$f(x) = \begin{cases} \frac{1}{2}x + \frac{1}{2} & 0 \leq x < 1 \\ \sqrt{1-(x-1)^2} & 1 \leq x \leq 2 \end{cases}$$

A. $1 + \dfrac{\pi}{2}$

B. $1 + \dfrac{\pi}{4}$

C. $\dfrac{3}{4} + \dfrac{\pi}{2}$

D. $\dfrac{3}{4} + \dfrac{\pi}{4}$

E. $\dfrac{3}{2}$

The answer is D.
A graph of the region is shown below. It has been divided into three subregions: a triangle, a rectangle, and a quarter circle. The areas of these subregions have been labeled $A_1, A_2,$ and A_3, respectively. The area A of the region is the sum of these three areas.

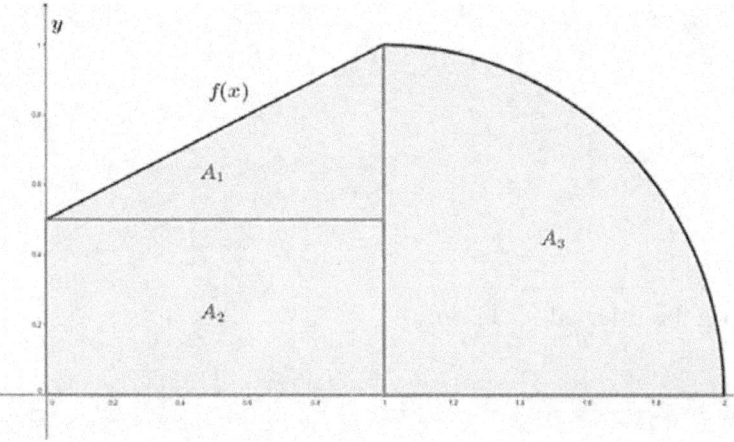

Rather than integrating, we will find A using basic geometrical formulas.

$$A = A_1 + A_2 + A_3$$
$$A = \frac{1}{2}(1)\left(\frac{1}{2}\right) + (1)\left(\frac{1}{2}\right) + \frac{1}{4}\pi(1)^2$$
$$A = \frac{3}{4} + \frac{\pi}{4}$$

CALCULUS

33. **The displacement s of a particle at time $t \geq 0$ is given by the following function.**

$$s(t) = \frac{1}{20}t^5 - \frac{5}{12}t^4 + \frac{4}{3}t^3 - 2t^2$$

Assume that all quantities are in SI units. For what values of t (in seconds) is the acceleration of the particle negative?

A. (0,1) only

B. (0,3) only

C. (1,2) only

D. (0,1) ∪ (1,2)

E. (0,1) ∪ (1,3)

The answer is A.
The acceleration a of the particle is given as a function of time t by $a(t) = s''(t)$. So the acceleration is negative whenever the second derivative of the displacement with respect to time is negative.

$a(t) = s''(t)$
$a(t) = t^3 - 5t^2 + 8t - 4$

A plot of $a(t)$ for $t \geq 0$ is given below.

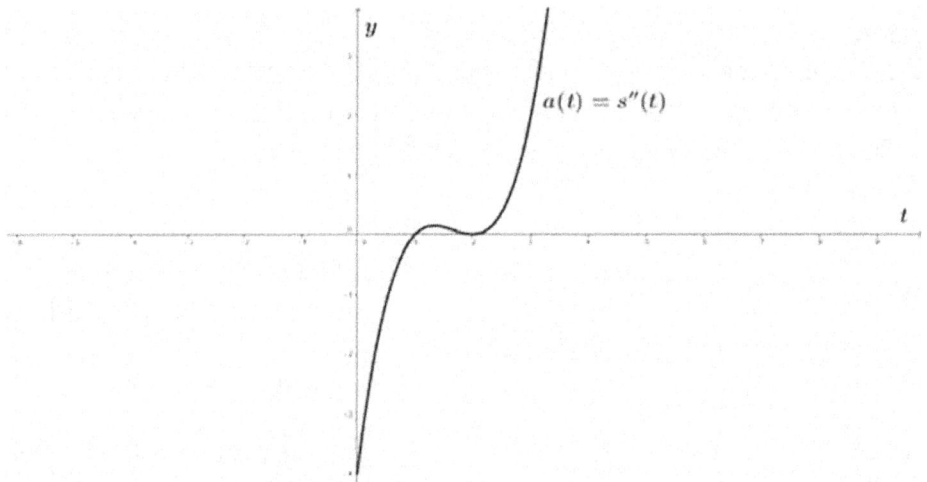

We can see from the graph that $a(t) < 0$ on the interval (0,1) only.

34. Use four trapezoids to estimate the area of the region bounded by the graph of $f(x) = \frac{1}{x^4+1}$ and the lines $x = 0, x = 2$, and $y = 0$. Round your answer to the nearest hundredth.

The answer is 1.07.

The region and the four trapezoids are sketched below.

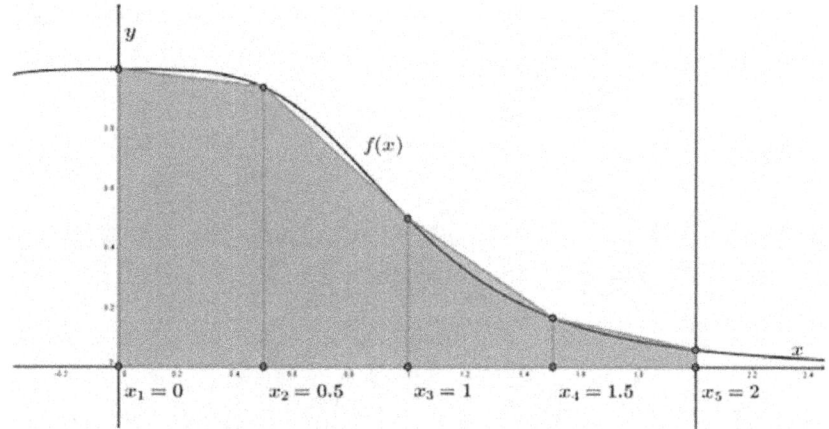

As you can see, the interval $[0,2]$ is partitioned into four subintervals by the points $x_1 = 0, x_2 = 0.5, x_3 = 1, x_4 = 1.5$, and $x_5 = 2$. We estimate the area A of the region using the Trapezoidal Rule below.

$$A \approx \frac{2-0}{2(4)}[f(0) + 2f(0.5) + 2f(1) + 2f(1.5) + f(2)]$$

$$A \approx \frac{1}{4}\left[1 + 2\left(\frac{16}{17}\right) + 2\left(\frac{1}{2}\right) + 2\left(\frac{16}{97}\right) + \frac{1}{17}\right]$$

$$A \approx 1.07$$

35. $\lim_{x \to 0^+} x^{\sin(x)} =$

   A. -1

   B. 0

   C. $\frac{1}{2}$

   D. 1

   E. 2

The answer is D.
The graph of the function $f(x) = x^{\sin(x)}$ is shown below.

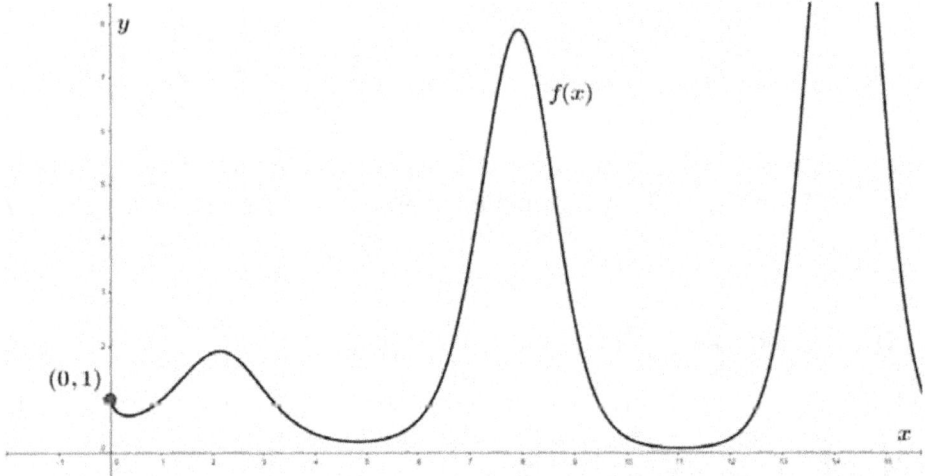

It certainly *looks* like the limit is 1. The limit is certainly not $-1, 0, \frac{1}{2}$, or 2 so eliminate A, B, C, and E.

CALCULUS

36. The acceleration a (in $\frac{m}{s^2}$) of a particle at time $t \geq 0$ (in s) is given by $a(t) = te^{-t^2}$. At $t = 0$ s, the velocity of the particle is $2.5 \frac{m}{s}$. What is the velocity (in $\frac{m}{s}$) of the particle at $t = 2$ s?

A. $2 - \frac{1}{2e^4}$

B. $2 + \frac{1}{2e^4}$

C. $3 - \frac{1}{2e^4}$

D. 3

E. $3 + \frac{1}{2e^4}$

The answer is C.
Find $v(t)$ by integrating $a(t)$ once with respect to t.

$v(t) = \int a(t) \, dt$

$v(t) = \int te^{-t^2} dt$ Let $u = -t^2$. Then $du = -2t \, dt \Rightarrow t \, dt = -\frac{1}{2} du$. We will continue to refer to the velocity as $v(t)$, despite the fact that we will temporarily change the variable to u.

$v(t) = -\frac{1}{2} \int e^u du$

$v(t) = -\frac{1}{2} e^u + C$

$v(t) = -\frac{1}{2} e^{-t^2} + C$ Next we'll find C by using the fact that the velocity is $2.5 \frac{m}{s}$ at $t = 0$ s.

$v(0) = 2.5$

$-\frac{1}{2} e^0 + C = 2.5$

$-0.5 + C = 2.5$

$C = 3 \Rightarrow v(t) = -\frac{1}{2} e^{-t^2} + 3$ So the velocity at $t = 2$ s is found as follows.

$v(2) = -\frac{1}{2} e^{-2^2} + 3$

$v(2) = -\frac{1}{2e^4} + 3$

CALCULUS

37. Find the equation of the line that passes through the point $(3, 4)$ and that, together with the $x-$ and $y-$axes, forms a triangular region in the first quadrant of minimum area.

   A. $3x + 4y - 25 = 0$

   B. $3x - 4y + 7 = 0$

   C. $4x + 3y - 24 = 0$

   D. $4x - 3y = 0$

   E. $x + y - 7 = 0$

The answer is C.
The point (1,2) satisfies all of the equations in the answer choices. However, the graphs of equations B and D do not enclose triangular regions in the first quadrant, as you can see below.

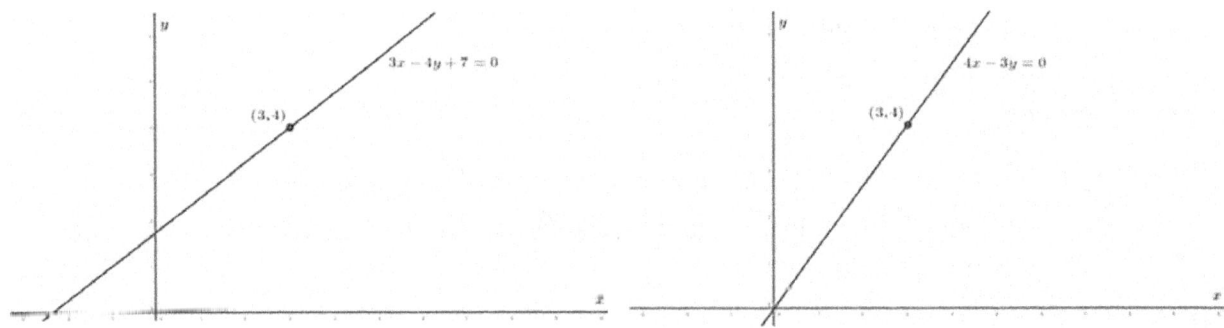

Eliminate B and D. Next we will show the graphs of the other three equations and compute the areas of their corresponding triangular regions in the first quadrant. Rather than integrating, we'll use the formula for the area of a triangle: $A = \frac{1}{2} bh$.

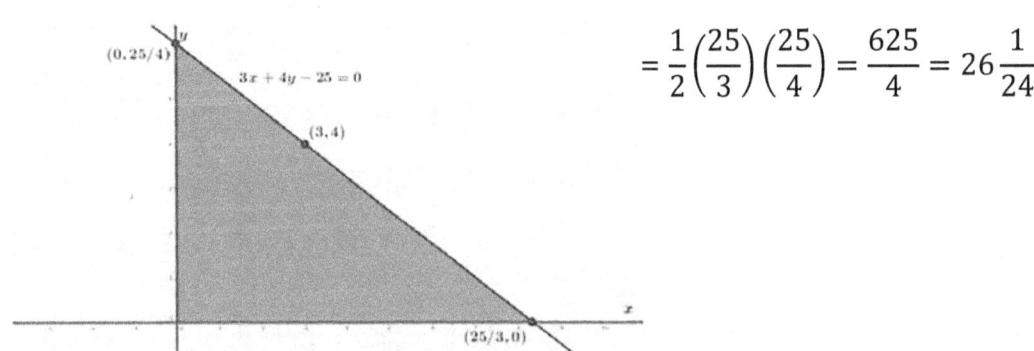

$$= \frac{1}{2}\left(\frac{25}{3}\right)\left(\frac{25}{4}\right) = \frac{625}{24} = 26\frac{1}{24}$$

CALCULUS

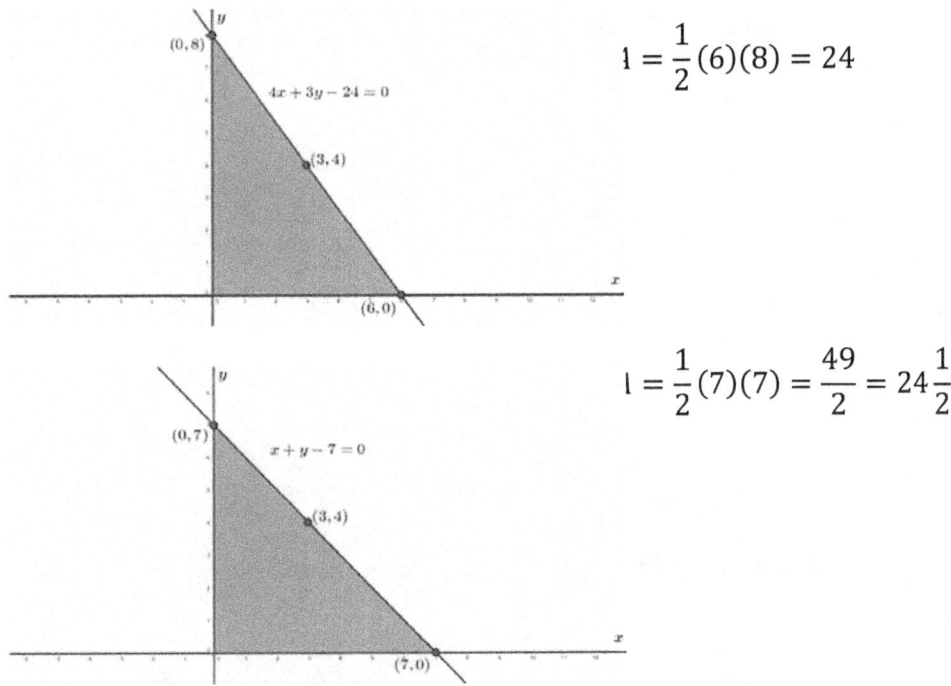

$$A = \frac{1}{2}(6)(8) = 24$$

$$A = \frac{1}{2}(7)(7) = \frac{49}{2} = 24\frac{1}{2}$$

The smallest of these areas is 24, so eliminate A and E, and choose C.
Note: We could have also solved this as an optimization problem using differential calculus.

38. Let $A = \int_{-1/2}^{1/2} f(x)\,dx$, $B = \int_{-1/2}^{1/2} g(x)\,dx$, and $C = \int_{-1/2}^{1/2} h(x)\,dx$, where $f(x) = \frac{1}{2(x^2+1)}$, $g(x) = \frac{x^2}{x^2+1}$, and $h(x) = e^{-x^2}$. Which of the following inequalities is true?

A. $A < B < C$

B. $A < C < B$

C. $B < A < C$

D. $B < C < A$

E. $C < A < B$

The answer is C.
All three functions are graphed below.

197

CALCULUS

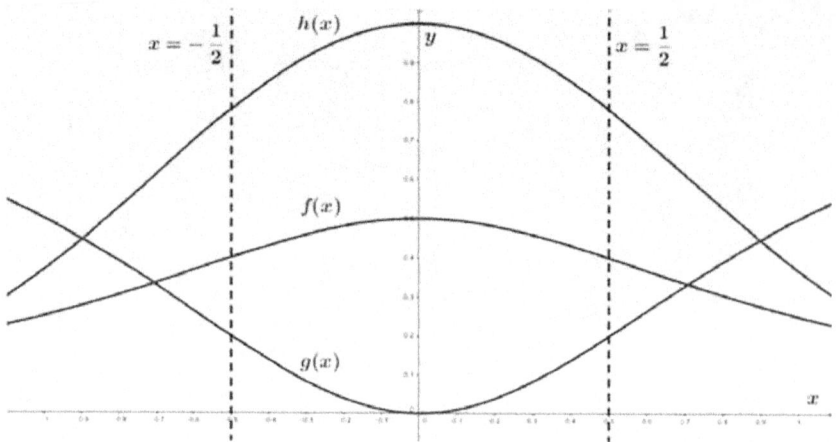

Since, for all x in the interval $\left[-\frac{1}{2}, \frac{1}{2}\right]$, $g(x) < f(x) < h(x)$, it follows that

$\int_{-1/2}^{1/2} g(x)\, dx < \int_{-1/2}^{1/2} f(x)\, dx < \int_{-1/2}^{1/2} h(x)\, dx$, Or $B < A < C$.

39. Let $f(x) = -\frac{1}{3}x^3 - x + 1$. Find any values of x in the interval $[-3, 3]$ at which the instantaneous rate of change of f equals the average rate of change of f on $[-3, 3]$.

   A. -3

   B. $\pm\sqrt{5}$

   C. $\pm\sqrt{3}$

   D. $\pm\sqrt{2}$

   E. 3

The answer is C.
We will equate $f'(x)$ to the average rate of change of f on $[-3,3]$, and then solve for the desired values of x.

$$f'(x) = \frac{f(3) - f(-3)}{3 - (-3)}$$
$$-x^2 - 1 = \frac{-11 - 13}{3 - (-3)}$$
$$-x^2 - 1 = -4$$
$$x^2 = 3$$
$$x = \pm\sqrt{3}$$

The graph of f is shown below. The secant line on $[-3,3]$ is shown in green, and the tangent lines at $x = \pm\sqrt{3}$ are shown in red. You can see that these three lines are all parallel, indicating

198

that the average rate of change of f on $[-3,3]$ equals the instantaneous rate of change of f at $x = \pm\sqrt{3}$.

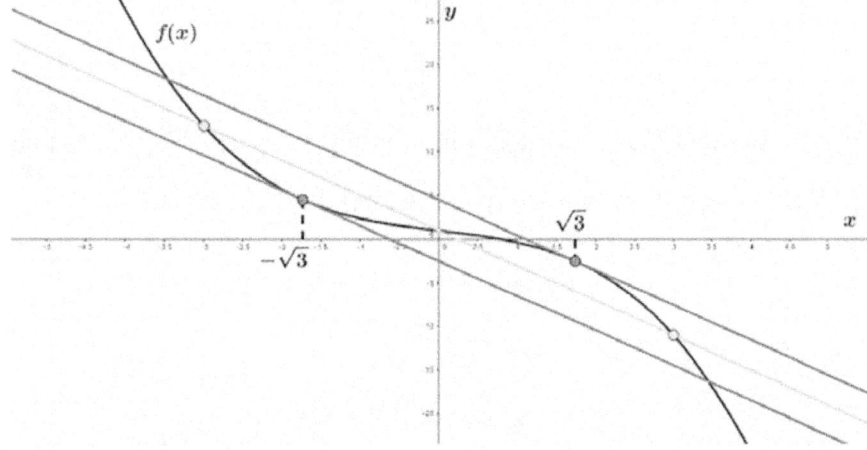

40. Consider the region bounded by the graph of $y = 4 - 2x^2$ and the x-axis. Find the area of the largest isosceles triangle that can be inscribed in this region with one vertex at the origin and the base parallel to the x-axis. Round your answer to the nearest hundredth.

The answer is 2.18
Begin by sketching the region with an isosceles triangle inscribed in it.

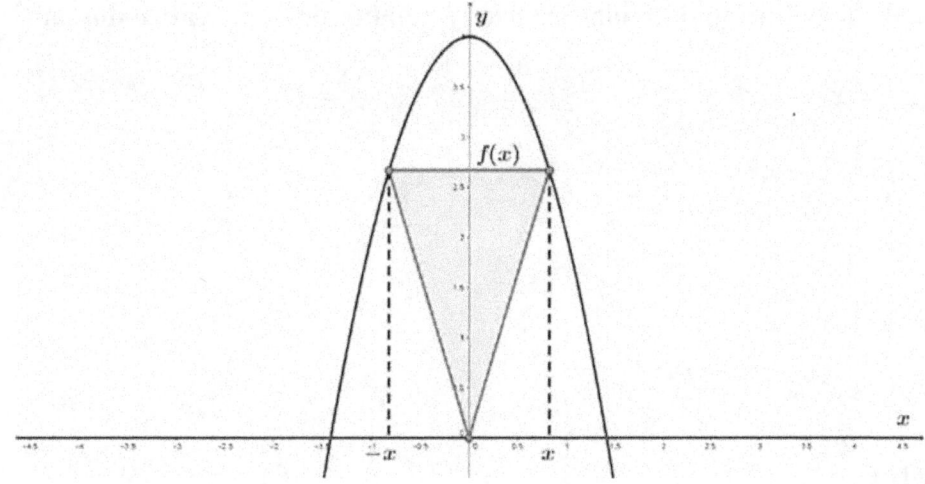

Letting $x > 0$, the base of the isosceles triangle is $x - (-x) = 2x$, and its height is $f(x)$. So we can write down the area A of the triangle as a function of x as follows.

$A(x) = \dfrac{1}{2} \cdot 2x \cdot f(x)$
$A(x) = x(4 - 2x^2)$
$A(x) = 4x - x^3$ Next we find the critical numbers of A.
$A'(x) = 8 - 6x^2 = 0$

$$x = \pm\sqrt{\frac{2}{3}}$$

Since we declared earlier that $x > 0$, we will only consider $x = \sqrt{\frac{2}{3}}$. To ensure that this critical number corresponds to a *minimum*, we apply the second derivative test.

$A''(x) = -12x$

$$A''\left(\sqrt{\frac{2}{3}}\right) = -12\sqrt{\frac{2}{3}} < 0$$

Since $A'' < 0$ at the critical number, the graph of A is *concave down* at that point, and so A does indeed have a relative minimum at this point. The minimum area is calculated below.

$$A\left(\sqrt{\frac{2}{3}}\right) = 4\sqrt{\frac{2}{3}} - 2\left(\sqrt{\frac{2}{3}}\right)^3 \approx 2.18$$

41. Which of the following functions is strictly monotonic on its entire domain?

   A. $f(x) = x^3 - 2x$

   B. $f(x) = x^3 - x$

   C. $f(x) = x^3 + x$

   D. $f(x) = x^2 - x$

   E. $f(x) = x^2 + x$

The answer is C.
A function f is strictly monotonic on its entire domain if f' does not change sign at any point in the domain of f. This guarantees that f is *either* strictly increasing *or* strictly decreasing on its entire domain. We will check the derivative of each function below.

(A) $f'(x) = 3x^2 - 2 \Rightarrow$ Changes sign at $x = \pm\sqrt{\frac{2}{3}} \Rightarrow$ Not monotonic

(B) $f'(x) = 3x^2 - 1 \Rightarrow$ Changes sign at $x = \pm\sqrt{\frac{1}{3}} \Rightarrow$ Not monotonic

(C) $f'(x) = 3x^2 + 1 \Rightarrow$ Positive for all real numbers $x \Rightarrow$ Strictly monotonic

(D) $f'(x) = 2x - 1 \Rightarrow$ Changes sign at $x = \frac{1}{2} \Rightarrow$ Not monotonic

(E) $f'(x) = 2x + 1 \Rightarrow$ Changes sign at $x = -\frac{1}{2} \Rightarrow$ Not monotonic

The graph of the function in answer choice C is increasing for all real x, as you can see from the figure below.

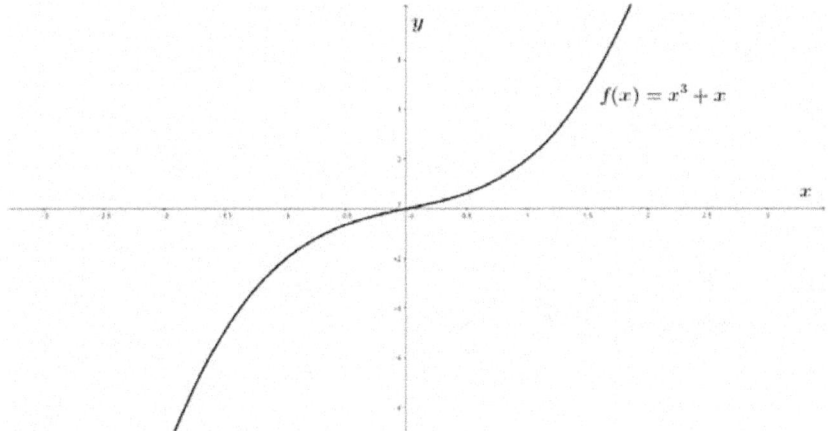

42. The radius of a circle is increasing at a constant rate of 3 $\frac{cm}{s}$. At what rate (in $\frac{cm^2}{s}$) is the area of the circle increasing when the radius is 5 cm?

   A. 6π

   B. 9π

   C. 10π

   D. 25π

   E. 30π

The answer is E.
The area A of a circle is related to the radius r of the circle by $A = \pi r^2$. Assuming that both A and r are implicit, differentiable functions of time, we can differentiate as follows.
$$\frac{dA}{dt} = 2\pi r \frac{dr}{dt}$$
Since $\frac{dr}{dt} = 3 \frac{cm}{s}$ was given, we have the following.
$$\frac{dA}{dt} = 2\pi r(3)$$
$\frac{dA}{dt} = 6\pi r$ Now we evaluate $\frac{dA}{dt}$ at $r = 5$ cm as follows.
$$\frac{dA}{dt}\bigg|_{r=5} = 6\pi(5) \quad so, \quad \frac{dA}{dt}\bigg|_{r=5} = 30\pi \frac{cm^2}{s}$$

43. $\lim\limits_{x \to \pi/2} \sec(x)\tan(x) =$

   A. $-\infty$

   B. 0

   C. 1

   D. ∞

   E. The limit does not exist.

The answer is D.
We begin by sketching the graph of $f(x) = \sec(x)\tan(x)$.

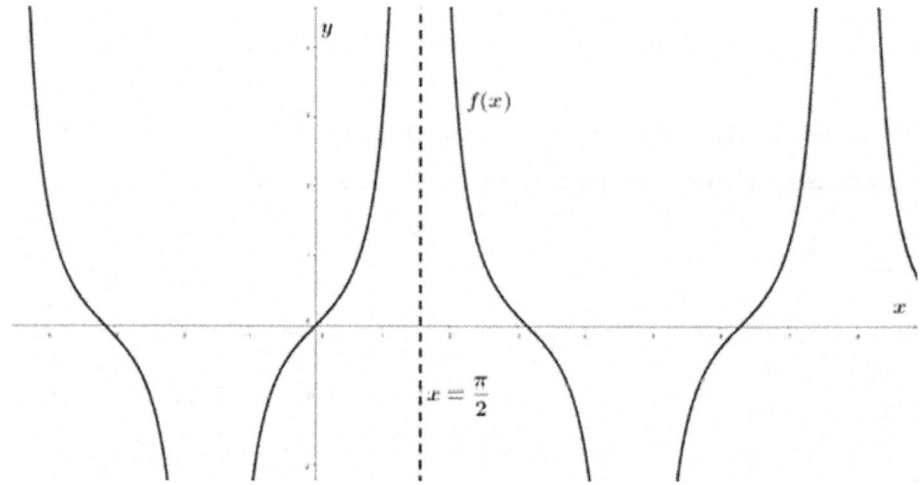

From the graph, we can see that $f(x) \to \infty$ as $x \to \dfrac{\pi}{2}$ from the left or the right.

CALCULUS

44. Find all real numbers c in $[a, b]$ such that $\int_a^b f(x)\, dx = f(c)(b - a)$ if $a = 0, b = 2$, and $f(x) = 3x^2 + 1$.

A. $\pm \frac{\sqrt{3}}{3}$

B. $\pm \frac{2\sqrt{3}}{3}$

C. $\pm \frac{4}{3}$

D. $\frac{2\sqrt{3}}{3}$

E. $\frac{4}{3}$

The answer is D.

$$\int_0^2 (3x^2 + 1)\, dx = (3c^2 + 1)(2 - 0)$$

$(x^3 + x)\big|_0^2 = 6c^2 + 2$

$(2^3 + 2) - (0^3 + 0) = 6c^2 + 2$

$10 = 6c^2 + 2$

$c^2 = \frac{4}{3}$

$c = \pm \sqrt{\frac{4}{3}} = \pm \frac{2\sqrt{3}}{3}$

Since $-\frac{2\sqrt{3}}{3}$ is not in the interval $[0,2]$, we discard it. Thus, the only solution is $c = \frac{2\sqrt{3}}{3}$.

XAMonline
The CLEP Specialist

Individual Sample Tests in eBook format with full explanations

eBooks

All 33 CLEP sample tests are available as eBook downloads from retail websites such as **Amazon.com** and **Barnesandnoble.com**

Title	ISBN
American Government	9781607875130
American Literature	9781607875079
Analyzing and Interpreting Literature	9781607875086
Biology	9781607875222
Calculus	9781607875376
Chemistry	9781607875239
College Algebra	9781607875215
College Composition	9781607875109
College Composition Modular	9781607875437
College Mathematics	9781607875246
English Literature	9781607875093
Financial Accounting	9781607875383
French	9781607875123
German	9781607875369
History of the United States I	9781607875178
History of the United States II	9781607875185
Human Growth and Development	9781607875444
Humanities	9781607875147
Information Systems	9781607875390
Introduction to Educational Psychology	9781607875451
Introductory Business Law	9781607875420
Introductory Psychology	9781607875154
Introductory Sociology	9781607875352
Natural Sciences	9781607875253
Precalculus	9781607875345
Principles of Macroeconomics	9781607875406
Principles of Microeconomics	9781607875468
Principles of Marketing	9781607875475
Principles of Management	9781607875468
Social Sciences and History	9781607875161
Spanish	9781607875116
Western Civilization I	9781607875192
Western Civilization II	9781607875208

TO ORDER Individual full length sample test are available from amazon or BARNES & NOBLE BOOKSELLERS

XAMonline
CLEP
Full Guides

TO ORDER — Complete study guides are available from amazon or BARNES & NOBLE BOOKSELLERS

CLEP College Algebra
ISBN: 9781607875307
Price: $34.99

CLEP Biology
ISBN: 9781607875314
Price: $34.99

CLEP Analyzing and
Interpreting Literature
ISBN: 9781607875260
Price: $34.99

CLEP College Composition
and Modular
ISBN: 9781607875277
Price: $16.99

CLEP College Mathematics
ISBN: 9781607875321
Price: $34.99

CLEP Psychology
ISBN: 9781607875291
Price: $34.99

CLEP Spanish
ISBN: 9781607875284
Price: $34.99

XAMonline
CLEP Subject Samplers

Collection by Topic
Sample Test Approach

CLEP Literature
ISBN: 9781607875833
Price: $24.99

CLEP Foreign Language
ISBN: 9781607875772
Price: $24.99

CLEP History
ISBN: 9781607875789
Price: $24.99

CLEP Sociology
ISBN: 9781607875796
Price: $24.99

CLEP Science
ISBN: 9781607875802
Price: $24.99

CLEP Mathematics
ISBN: 9781607875819
Price: $24.99

CLEP Business
ISBN: 9781607875826
Price: $24.99

TO ORDER — Complete sample tests are available from **amazon** or **BARNES & NOBLE** BOOKSELLERS

XAMonline
CLEP Favorites
Collection by Topic
Sample Test Approach

CLEP 5
ISBN: 9781607875307
Price: $25.99

CLEP Military Favorites
ISBN: 9781607875314
Price: $25.99

TO ORDER Complete sample tests are available from **amazon** or **BARNES & NOBLE** BOOKSELLERS

www.ingramcontent.com/pod-product-compliance
Lightning Source LLC
Chambersburg PA
CBHW080732230426
43665CB00020B/2712